Electric Wiring: Domestic

By the same author

17th Edition IEE Wiring Regulations: Design and Verification of Electrical Installations, ISBN 978-0-7506-8721-8

17th Edition IEE Wiring Regulations: Explained and Illustrated, ISBN 978-0-7506-8720-1

17th Edition IEE Wiring Regulations: Inspection, Testing and Certification, ISBN 978-0-7506-8719-5

PAT: Portable Appliance Testing, ISBN 978-0-7506-8736-2

Wiring Systems and Fault Finding, ISBN 978-0-7506-8734-8

Electrical Installation Work, ISBN 978-0-7506-8733-1

Electric Wiring: Domestic

Thirteenth edition

Brian Scaddan IEng, MIET

AMSTERDAM • BOSTON • HEIDELBERG • LONDON
NEW YORK • OXFORD • PARIS • SAN DIEGO
SAN FRANCISCO • SINGAPORE • SYDNEY • TOKYO
Newnes is an imprint of Elsevier

Newnes is an imprint of Elsevier
Linacre House, Jordan Hill, Oxford OX2 8DP, UK
30 Corporate Drive, Suite 400, Burlington, MA 01803, USA

First published 1940
Second edition 1944
Third edition 1947
Fourth edition 1954
Fifth edition 1957
Sixth edition 1967
Seventh edition 1969
Eighth edition 1983
Ninth edition 1989
Tenth edition 1992
Eleventh edition 1997
Twelfth edition 2003
Thirteenth edition 2008

British Library Cataloguing in Publication Data
Scaddan, Brian
   Electric wiring – domestic. – 13th ed.
   1. Electric wiring, Interior 2. Electric apparatus and appliances – Installation
   I. Title
   621.3'1924

Library of Congress Control Number: 2008927627
A catalog record for this book is available from the Library of Congress

ISBN: 978-0-7506-8735-5

For more information on all Newnes publications visit
our website at www.elsevierdirect.com

Typeset by Charon Tec Ltd., A Macmillan Company. (www.macmillansolutions.com)

Printed and bound in Slovenia

08 09 10 11   11 10 9 8 7 6 5 4 3 2 1

Contents

Preface

Electric Wiring: Domestic has for many years been acknowledged as the standard guide to the practical aspects of domestic electric wiring. It seeks to address the areas of most concern to the qualified electrician, especially design and testing. It will also be a useful addition to the resources available for students working towards NVQs or City & Guilds qualifications.

This book is also a vital reference source for many other professionals and operatives whose work demands a knowledge of electrical installations, including electrical engineers, heating engineers, architects and maintenance staff. The contents will be of value to those intending to gain a Domestic Installer Scheme Qualification which relates to Part 'P' of the Building Regulations. It is not intended as a DIY manual, although some non-qualified persons may find certain topics useful before calling in qualified operatives.

The contents of this new edition cover current professional best practice and are fully compliant with the 17th Edition IEE Wiring Regulations.

Brian Scaddan, April 2008

Material on Part P in Chapter 1 is taken from *Building Regulations Approved Document P: Electrical Safety – Duellings*, P1 Design and Installation of electrical installations (The Stationery Office, 2006) ISBN 9780117036536. © Crown copyright material is reproduced with the permission of the Controller of HMSO and Queen's Printer for Scotland.

Acknowledgements

I would like to thank Paul Clifford for his thorough technical proof reading.

CHAPTER 1

The UK Generation, Transmission and Distribution System

In the early days of electricity supply, each town or city in the United Kingdom had its own power station which supplied the needs of its particular area.

Standardization was not evident and many different voltages and frequencies were used throughout the country. By the time of the First World War (1914–1918), there were some 600 independent power stations in use. However, the heavy demands made by the war industry showed the inadequacies of the system and several select committees were set up to investigate possible changes. Little was achieved until 1926, when it was suggested that 126 of the largest and most efficient power stations should be selected and connected by a grid of high-voltage transmission lines covering the whole country, and, at the same time, the frequency standardized at 50 Hz. The remaining power stations would be closed down and local supply authorities would obtain their electricity in bulk from the grid, via suitable substations. The system voltage was 132 000 V (132 kV) and the supply frequency 50 Hz.

On 1st April 1948, the whole of the electricity supply industry was nationalized and, in 1957, the 'Central Authority' responsible for the generation of electricity was renamed the 'Central Electricity Generating Board' (CEGB).

Since then, of course, the electricity industry has become privatized and the CEGB has been replaced by the National Grid Company, which buys, at the lowest price, generated electricity from such

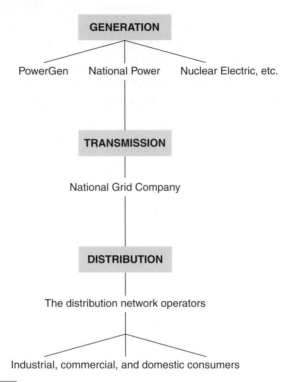

FIGURE 1.1

companies as National Power, PowerGen, Nuclear Electric, French Electrique and Scottish Hydro Electric.

Electricity boards have become Distribution Network Operators (DNOs) and they, in turn, buy electrical energy from the National Grid Company to distribute to their own consumers.

The broad structure of the industry is shown in Figure 1.1.

VOLTAGE BANDS

The very nature of the grid system is such that power has to be transmitted over large distances. This immediately creates a problem of voltage drop. To overcome this problem, a high voltage is used for transmission (400 or 132 kV), the 400 kV system being known as the 'Super Grid'. We cannot, however, generate at such high voltages (the maximum in modern generators is 25 kV) and transformers are used to step up the generated voltage to the transmission voltage. At the end of a transmission line is a grid substation, where the requirements of the grid system in that area can be controlled and where the transmission voltage is stepped down via a transformer to 132 kV.

It is at this stage that the different DNOs distribute the power required by their consumers around that particular area. The system voltage is then further reduced at substations to 33000, 11000 and 415/240 V.

The declared voltage at consumers' terminals is now 400 V three-phase/230 V single-phase +10 to 6%. However, the measured voltage is still likely to be 415/240 V for many years.

THE ELECTRICITY SAFETY, QUALITY AND CONTINUITY REGULATIONS 2002 (ESQCR)

Clearly, from the public's point of view, there must be safeguards from the dangers that electricity supply systems may create and also guarantees that, other than in exceptional circumstances, a constant supply will be maintained.

Such safeguards and guarantees are embedded in the Electricity Supply Regulations 1988.

With regards to a consumer installation, the regulations advise the supplier (usually the DNO) to provide a service cable, protective device and any wiring up to the consumer's supply terminals.

They also require the supplier to give written details of external loop impedance and prospective short-circuit values.

The regulations also permit the supplier to withhold or discontinue a supply if it is considered that the consumer's installation is not safe or could interfere with the public supply.

THE ELECTRICITY AT WORK REGULATIONS 1989

These regulations come under the umbrella of the Health and Safety at Work Act 1974 and supersede the Factories Act (Special Regulations) 1908 and 1944. They affect **every** person at work, whether an employer, an employee or a self-employed person, and they place a duty on each person to carry out safe working practices with regards to matters (electrical) that are within his or her control.

This piece of legislation is the only one in the United Kingdom that assumes anyone who contravenes certain regulations to be **guilty**. Such a person then has to demonstrate that he or she took all reasonable steps to prevent danger in order to prove his or her innocence and avoid prosecution (Figure 1.2).

THE IEE WIRING REGULATIONS (BS 7671)

These regulations, although having British Standard status, are non-statutory. However, they may be used in a court of law to claim compliance with statutory regulations such as the Electricity at Work Regulations 1989.

They are, basically, a set of recommendations dealing with the safe design, construction, inspection and testing of low-voltage installations. There is no point at this stage in expanding further on these regulations, as they will be referred to throughout the remaining chapters.

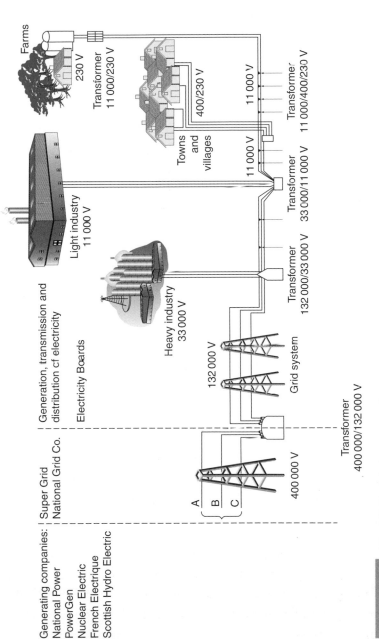

Generating companies:
National Power
PowerGen
Nuclear Electric
French Electrique
Scottish Hydro Electric

Super Grid
National Grid Co.

Generation, transmission and
distribution of electricity

Electricity Boards

Farms

230 V

Transformer
11 000/230 V

400/230 V

Towns
and
villages

11 000 V

11 000 V

Transformer
11 000/400/230 V

Light industry
11 000 V

Transformer
33 000/11 000 V

Heavy industry
33 000 V

Transformer
132 000/33 000 V

132 000 V

Grid system

400 000 V

A
B
C

Transformer
400 000/132 000 V

FIGURE 1.2   The UK Electrical energy system.

THE BUILDING REGULATIONS 2000 PART 'P'

The following material is taken from *The Building Regulations 2000 approved document* P. © Crown Copyright material is reproduced with the permission of the controller of HMSO and Queen's Printer for Scotland.

Part 'P'

Part 'P' of the Building Regulations requires that certain electrical installation work in domestic dwellings be certified and notified to the Local Authority Building Control (LABC). Failure to provide this notification may result in substantial fines.

Some approval bodies offer registration for **all** electrical work in domestic premises, these are known as full scope schemes (FS); other bodies offer registration for **certain limited** work in special locations such as kitchens, bathrooms, gardens, etc., these are known as defined scope schemes (DS).

In order to achieve and maintain competent person status, all approval bodies require an initial and thereafter annual registration fee and inspection visit.

Approval bodies (full scope FS and defined scope DS)

NICEIC	(FS) & (DS)	0870 013 0900
NAPIT	(FS) & (DS)	0870 444 1392
ELESCA	(FS) & (DS)	0870 749 0080
BSI	(FS)	01442 230 442
BRE	(FS)	0870 609 6093
CORGI	(DS)	01256 392 200
OFTEC	(DS)	0845 658 5080

Certification of notifiable work

a. Where the installer is registered with a Part P competent person self-certification scheme

1.18 Installers registered with a Part P competent person self-certification scheme are qualified to complete BS 7671 installation certificates and should do so in respect of every job they undertake. A copy of the certificate should always be given to the person ordering the electrical installation work.

1.19 Where installers registered with Part P competent person self-certification scheme, a Building Regulations compliance certificate must be issued to the occupant either by the installer or the installer's registration body within 30 days of the work being completed. The relevant building control body should also receive a copy of the information on the certificate within 30 days.

1.20 The Regulations call for the Building Regulations compliance certificate to be issued to the occupier. However, in the case of rented properties, the certificate may be sent to the person ordering the work with a copy sent also to the occupant.

b. Where the installer is *not* registered with a Part P competent person self-certification scheme but qualified to complete BS 7671 installation certificates

1.21 Where notifiable electrical installer work is carried out by a person **not** registered with a Part P competent person self-certification the work should be notified to a building control body (the local authority or an approved inspector) before work starts. Where the work is necessary because of an emergency the building control body should be notified as soon as possible. The building control body becomes responsible for making sure the work is safe and complies with all relevant requirements of the Building Regulations.

1.22 Where installers are qualified to carry out inspection and testing and completing the appropriate BS 7671 installation certificate, they should do so. A copy of the certificate should then be given to the building control body. The building control body will take this certificate into account in deciding what further action (if any) needs to be taken to make sure that the work is safe and complies fully with all relevant requirements. Building control bodies may ask for evidence that installers are qualified in this case.

1.23 Where the building control body decides that the work is safe and meets all building regulation requirements it will issue a building regulation completion certificate (the local authority) on request or a final certificate (an approved inspector).

c. Where installers are not qualified to complete BS 7671 completion certificates

1.24 Where such installers (who may be contractors or DIYers) carry out notifiable electrical work, the building control body must be notified before the work starts. Where the work is necessary because of an emergency the building control body should be notified as soon as possible. The building control body then becomes responsible for making sure that the work is safe and complies with all relevant requirements in the Building Regulations.

1.25 The amount of inspection and testing needed is for the building control body to decide based on the nature and extent of the electrical work. For relatively simple notifiable jobs, such as adding a socket-outlet to a kitchen circuit, the inspection and testing requirements will be minimal. For a house re-wire, a full set of inspection and tests may need to be carried out.

1.26 The building control body may choose to carry out the inspection and testing itself, or to contract out some or all of the work to a special body which will then carry out the work on its behalf. Building control bodies will carry out the necessary inspection and testing at their expense, not at the householders' expense.

1.27 A building control body will **not** issue a BS 7671 installation certificate (as these can be issued only by those carrying out the work), but only a Building Regulations completion certificate (the local authority) or a final certificate (an approved inspector).

Third party certification

1.28 Unregistered installers should not themselves arrange for a third party to carry out final inspection and testing. The third party – not having supervised the work from the outset – would not be in a position to verify that the installation work complied fully with BS 7671:2008 requirements. An electrical installation certificate can be **issued only by the installer** responsible for the installation work.

1.29 A third party could only sign a BS 7671:2008 Periodic Inspection Report or similar. The Report would indicate that electrical safety tests had been carried out on the installation which met BS 7671:2008 criteria, but it could not verify that the installation complied fully with BS 7671:2008 requirements – for example with regard to routing of hidden cables.

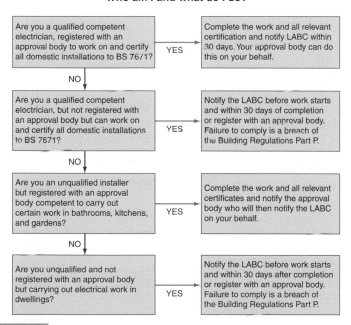

Who am I and what do I do?

| Are you a qualified competent electrician, registered with an approval body to work on and certify all domestic installations to BS 7671? | YES | Complete the work and all relevant certification and notify LABC within 30 days. Your approval body can do this on your behalf. |

NO

| Are you a qualified competent electrician, but not registered with an approval body but can work on and certify all domestic installations to BS 7671? | YES | Notify the LABC before work starts and within 30 days of completion or register with an approval body. Failure to comply is a breach of the Building Regulations Part P. |

NO

| Are you an unqualified installer but registered with an approval body competent to carry out certain work in bathrooms, kitchens, and gardens? | YES | Complete the work and all relevant certificates and notify the approval body who will then notify the LABC on your behalf. |

NO

| Are you unqualified and not registered with an approval body but carrying out electrical work in dwellings? | YES | Notify the LABC before work starts and within 30 days after completion or register with an approval body. Failure to comply is a breach of the Building Regulations Part P. |

FIGURE 1.3

Table 1.1 Examples of Work Notifiable and Not Notifiable.

Notifiable (YES)　　　　Not Notifiable (NO)　　　　Not Applicable (N/A)

EXAMPLES OF WORK	LOCATION A Within Kitchens, Bath/Shower Room, Gardens, Swimming/ Paddling Pools & Hot Air Saunas	LOCATION B Outside of Location A
A complete new installation or rewire	YES	YES
Consumer unit change	YES	YES
Installing a new final circuit (e.g. for lighting, socket outlets, a shower or a cooker)	YES	YES
Fitting and connecting an electric shower to an existing wiring point	YES	N/A
Adding a socket outlet to an existing final circuit	YES	NO
Adding a lighting point to an existing final circuit	YES	NO
Adding a fused connection unit to an existing final circuit	YES	NO
Installing and fitting a storage heater including final circuit	YES	YES
Installing extra low-voltage lighting (other than pre-assembled CE marked sets)	YES	YES
Installing a new supply to a garden shed or other building	YES	N/A
Installing a socket outlet or lighting point in a garden shed or other detached outbuilding	YES	N/A
Installing a garden pond pump, including supply	YES	N/A
Installing an electric hot air sauna	YES	N/A
Installing a solar photovoltaic power supply	YES	YES
Installing electric ceiling or floor heating	YES	YES
Installing an electricity generator	YES	YES
Installing telephone or extra low-voltage wiring and equipment for communications, information technology, signalling, control or similar purposes	YES	NO

(*continued*)

Table 1.1 *Continued*

Notifiable (YES)	Not Notifiable (NO)	Not Applicable (N/A)
EXAMPLES OF WORK	**LOCATION A** Within Kitchens, Bath/Shower Room, Gardens, Swimming/ Paddling Pools & Hot Air Saunas	**LOCATION B** Outside of Location A
Installing a socket outlet or lighting point outdoors	YES	YES
Installing or upgrading main or supplementary equipotential bonding	NO	NO
Connecting a cooker to an existing connection unit	NO	NO
Replacing a damaged cable for a single circuit, on a like-for-like basis	NO	NO
Replacing a damaged accessory, such as a socket outlet	NO	NO
Replacing a lighting fitting	NO	NO
Providing mechanical protection to an existing fixed installation	NO	NO
Fitting and final connection of storage heater to an existing adjacent wiring point	NO	NO
Connecting an item of equipment to an existing adjacent connection point	NO	NO
Replacing an immersion heater	NO	NO
Installing an additional socket outlet in a motor caravan	N/A	N/A

EARTHING SYSTEMS

The UK electricity system is an earthed system, which means that the star or neutral point of the secondary side of distribution transformers is connected to the general mass of earth.

In this way, the star point is maintained at or about 0 V. Unfortunately, this also means that persons or livestock in contact with a live part and earth are at risk of electric shock (Figure 1.4).

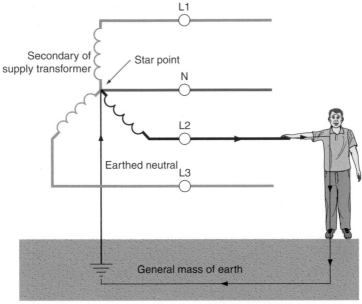

Shock current flows through person, through mass of earth to star point through L2 line winding back to person

FIGURE 1.4   Shock Path.

There are, however, methods of reducing the shock risk and these will be discussed in Chapter 3.

There are three main methods of earthing used in the United Kingdom, these are the TT system, the TN-S system, and the TN-C-S system. The letter T is the first letter of the French word for earth 'terre', and indicates a direct contact to the general mass of earth. The letter N indicates that there is also the connection of a conductor to the star or neutral point of the supply transformer, which is continuous throughout the distribution system and terminates at the consumer's intake position. The letters C and S mean 'combined' and 'separate', respectively.

So a TT system has the star or neutral point of the supply transformer directly connected to earth by means of an earth electrode, and the earthing of the consumer's installation is also directly connected to earth via an earth electrode (Figure 1.5). This system is typical of an overhead line supply in a rural area.

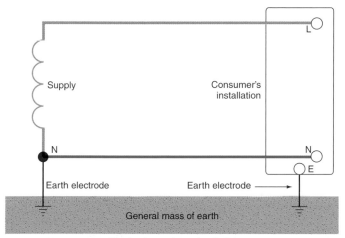

(Single-phase shown for clarity)

FIGURE 1.5 TT system.

A TN-S system has the star point of the supply transformer connected to earth. Also the outer metallic sheaths of the distribution cable and, ultimately, the service cable are also connected to the star point. Hence, there are **separate** (S) metallic earth and neutral conductors throughout the whole system (Figure 1.6).

A TN-C-S system (also known as protective multiple earthing, PME) has the usual star connection to earth and the metallic sheaths of the distribution and service cables also connected to the star point. In this case, however, the outer cable sheath is also used as a neutral conductor (i.e. it is a combined (C) earth and neutral).

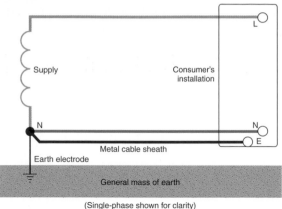

(Single-phase shown for clarity)

Separate neutral and earth conductors

FIGURE 1.6   TN-S system.

However, the system inside the consumer's premises continues to have separate (S) earth and neutral conductors (Figure 1.7).

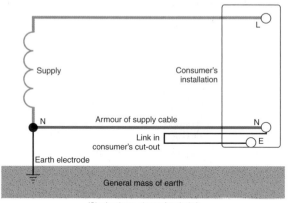

(Single-phase shown for clarity)

Combined earth and neutral conductor for supply.
Separate earth and neutral in consumer's installation

FIGURE 1.7   TN-C-S system.

These are the three main earthing systems used in the United Kingdom. They all rely on an earthed star point of the supply transformer and various methods of providing an earth path for fault currents.

Domestic Electrical Installations

THE MAIN INTAKE POSITION

Unless domestic premises are extremely large, it is unlikely that a three-phase supply would be needed, and consequently only single-phase systems will be considered here. Figures 2.1, 2.2 and 2.3 illustrate the typical intake arrangements for TT, TN-S and TN-C-S systems.

Although many TT installations are protected by one single 30 mA residual current device (RCD) (as shown in Figure 2.1), this does

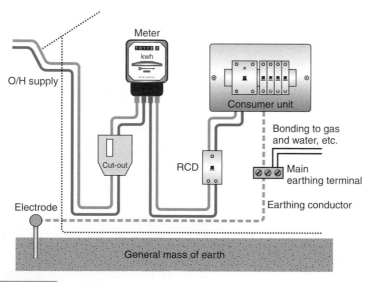

FIGURE 2.1 TT system.

15

not conform to the IEE Regulations regarding 'installation circuit arrangement'.

The requirement is that circuits which need to be separately controlled, for example lighting and power, remain energized in the event of the failure of any other circuit of the installation. Hence, an earth fault on, say, a socket-outlet circuit would cause the whole of the installation to be cut off if protected by one 30 mA RCD.

One preferred arrangement is to protect the whole installation by a 100 mA RCD and, using a 'split-load' consumer unit, protect the socket-outlet circuits with a 30 mA RCD (Figure 2.2).

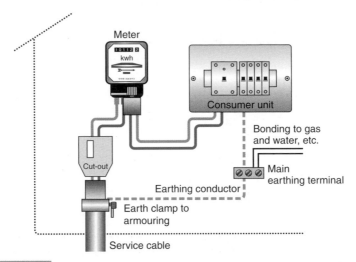

FIGURE 2.2   TN-S system.

Alternatively, combined RCD/CB devices (RCBOs) may be used to protect each circuit individually (Figure 2.3).

In many domestic situations, 'off-peak' electricity is used, as this can be a means of reducing electricity bills. Energy is consumed out of normal hours, for example 11.00 pm to 7.00 am, and the tariff (the charge per unit of energy used) is a lot less.

This arrangement lends itself to the use of storage heaters and water heating, and the supply intake equipment will incorporate

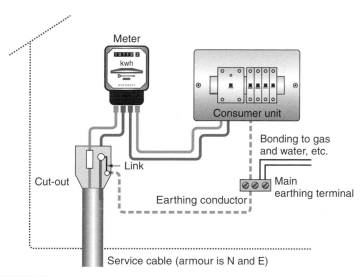

FIGURE 2.3 TN-C-S system.

special metering arrangements. The DNOs have their own variations on a common theme, depending on consumer's requirements, but, typically, the supply from the cut-out (this houses the DNOs fuse and neutral) feeds a digital meter from which three consumer units are fed: one for normal use, one for storage heating and one for water heating (Figure 2.4).

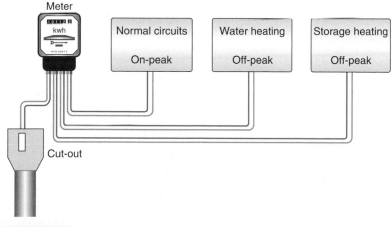

FIGURE 2.4 Off-peak system.

In these cases, these meters or telemeters, as they are known, are switched on and off by radio signals activated from the DNOs centre. In the case of water heating, there is normally a 'mid-day boost' for about 2 hours. In most DNO areas, electricity used during the night by normal circuits, that is lighting and power, attracts a lower tariff. As a result, it is cost-effective to carry out washing and drying activities during this nighttime period.

Many older installations incorporate the 'white meter' arrangement which uses a separate meter to register energy used during off-peak periods. Although most new installations are based around the tele-metering system, older metering installations are still valid. They are all variations on the same theme, that is they use electricity out-side normal hours and the charge per unit will be less.

The main earthing terminal

The intake arrangements shown in Figures 2.1–2.4 all indicate a main earthing terminal separate from the consumer unit. In fact, most modern units have an integral earth bar, which can accommodate all the circuit earths or circuit protective conductors (cpc's), the earthing conductor and the main protective bonding conductors. However, it is probably more convenient to have a separate main earthing termi-nal to which is connected the earthing conductor from the consumer unit, the earthing conductor to the means of earthing (earth electrode, cable sheath, etc.) and the main protective bonding conductors. This arrangement is particularly useful when an installation is under test.

As mentioned, the main earthing terminal is a point to which all main protective bonding conductors are connected. These conductors connect together gas, water and oil services etc., and in so doing, maintain such services within the premises at or about earth potential (i.e. 0 V). It must be remembered that bonding the installation earthing to these services is not done to gain an earth, many services are now run in non-metallic materials and it is within the premises that bonding is so very important. This aspect will be dealt with in greater detail in Chapter 3.

From a practical point of view, bonding of gas services should be carried out within 600 mm of the gas meter on the consumer's side of the meter, and as near as possible to the water intake position, once again on the consumer's side (Figure 2.5).

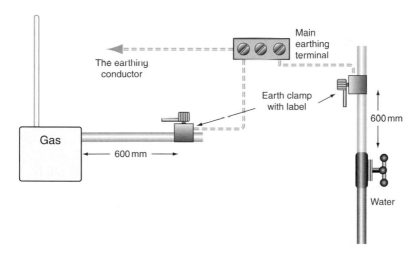

FIGURE 2.5 Bonding conductors.

There is no reason why bonding to main services should be carried out individually and separately, provided that the bonding conductors are unbroken. This will prevent conductors being accidentally pulled out of the bonding clamp terminal, leaving one or other of the services unbonded (Figure 2.6).

Main isolation

The main intake position houses, usually as part of the consumer unit, the means to isolate the supply to the whole installation, and there is a requirement to ensure that such isolation be accessible

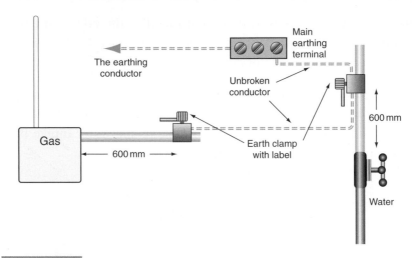

FIGURE 2.6   Bonding conductors.

at all times. So the means of isolation should not be housed in cupboards used for general household storage. Unfortunately, the design of many domestic premises tends to relegate this important equipment to areas out of sight and inaccessible to the occupier.

Earth electrodes and TT systems

A TT system requires an earth electrode at the consumer's premises. Such an electrode must be protected from corrosion and mechanical damage and the ideal arrangement is as shown in Figure 2.7.

CIRCUITS

Domestic circuits are either radial or ring final circuits and are likely to be arranged in the following ways:

- *Radial circuits* are fed from the consumer unit and run in either a chain or like the spokes of a wheel (i.e. they radiate out from their source). Typical of domestic radial circuits are lighting, water heating, storage heating and cooking.

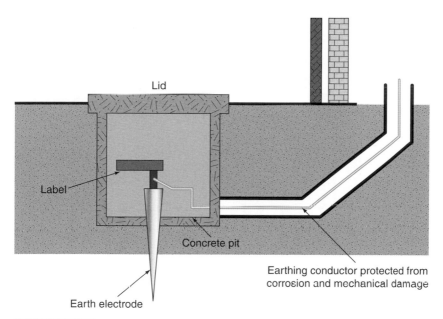

Lid

Label

Concrete pit

Earth electrode

Earthing conductor protected from
corrosion and mechanical damage

FIGURE 2.7 Earth electrode installation.

- *Ring final circuits* are almost unique to the United Kingdom.
  The ring final circuit is always used to feed 13 A socket outlets
  to BS 1363. The circuit starts at the consumer unit, loops in
  and out of each socket and, finally, returns to the consumer
  unit to terminate in the same terminals as it started.

Let us now look at the various circuits in a little more detail.

Lighting circuits

The 'loop-in' system

This is the most common of all lighting circuitry and, as the name
suggests, circuit cables simply 'loop' in and out of each lighting
point (Figure 2.8).

It should be noted that the cpc's have been omitted from Figure
2.8 for clarity. They must, however, always be present. In the case

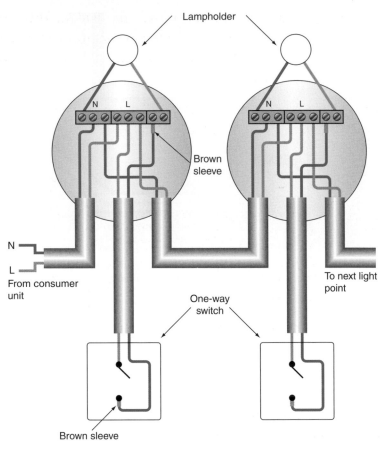

FIGURE 2.8   One way switching.

of flat twin with cpc cable (6242 Y), the bare cpc must be sheathed with green and yellow PVC sleeving, and the blue (old black) conductor from the switch to the light point must have some brown sleeving at both ends to indicate that it is a line conductor (not a neutral, as blue (old black) would signify). This conductor is known as a switch wire. If, however, the wiring system were formed using single conductors enclosed in conduit, then both conductors from the switch to the light point would be brown (old red) and no sleeving would be needed. For further details, see later sections in this chapter.

It is acceptable to wire two or more lighting points from the previous one (i.e. 3 and 4 from 2 in Figure 2.9).

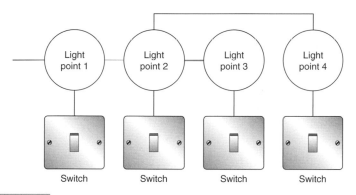

Lighting circuit with no junction box.

However, this causes problems of space in the ceiling rose and overcrowding of terminals. So it is perhaps best to avoid this practice and use a lighting junction box (Figure 2.10).

The need to wire in this way usually arises due to an addition to an existing system or, in the case of a new installation, when one point is very remote from the rest and when making it part of a continuous chain of points would result in considerable extra cable length (Figure 2.11).

There are many instances where two or more points are required to be controlled by one switch. In this case, each extra point is wired directly to the lampholder terminals of the previous point (Figure 2.12, see page 25).

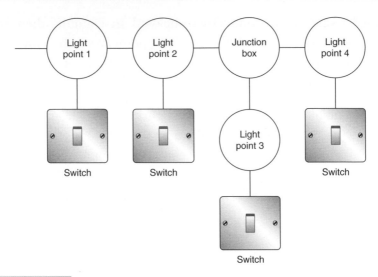

FIGURE 2.10 Lighting circuit with junction box.

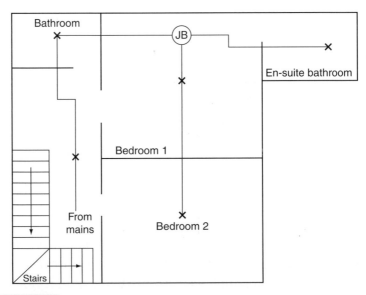

FIGURE 2.11 Lighting circuit with junction box.

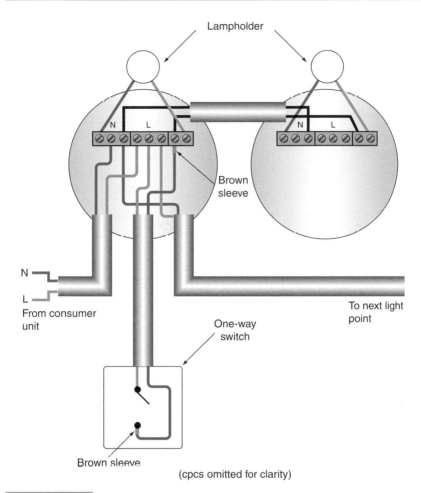

Lampholder

N   L   N   L

Brown
sleeve

N
L
From consumer
unit

To next light
point

One-way
switch

Brown sleeve

(cpcs omitted for clarity)

FIGURE 2.12 One way switch controlling two points.

So far, we have only considered one-way switching and, of course, most domestic premises have two-way systems and, in some cases, two-way and intermediate.

Two-way switching is typical of the control of lighting for stairwells. Two-way and intermediate switching is typical of the control of lighting for long corridors or three-storey dwellings with two or more landings, where lights need to be controlled from more than two places.

Two-way switching

There are two methods of wiring a two-way switching system. The first is achieved by running a three-core and cpc cable between two-way switches, via the lighting point (Figure 2.13).

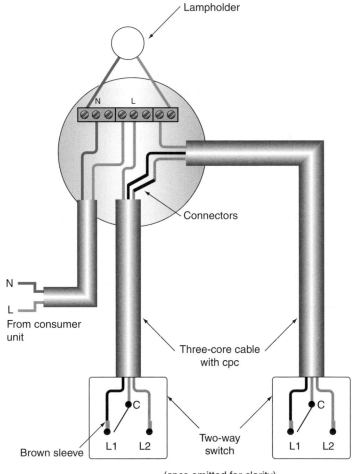

Lampholder

N   L

Connectors

N

L

From consumer unit

Three-core cable with cpc

Two-way switch

Brown sleeve   L1   L2   L1   L2

(cpcs omitted for clarity)

FIGURE 2.13   Two way switching via the ceiling rose.

The second method is to wire a twin cable to one two-way switch (as in one-way switching) and then to run a three-core cable from this to the other two-way switch (Figure 2.14).

The disadvantage with the first method is that the grey (old blue) and black (old yellow) conductors (found in three-core flat cable, 6243 Y), known as strappers, have to be terminated in the light point enclosure. As a result, connectors or crimps must be used as there are no terminals provided. This results in overcrowding of conductors in a confined space and as such this method is rarely used.

With the second method, the lighting point is not physically involved with the three-core cable. It should be noted that the grey (old blue) and black (old yellow) strappers are **line** conductors and must be identified as such with brown (old red) sleeving.

It is useful to note that the second method is a simple means of converting one-way switching to two-way, without disturbing the wiring to the light point. All that is required is a change of switch from one-way to two-way and the installation of an extra two-way switch with three-core cable wired between them. Typical of this would be converting a one-way system in, say, a bedroom, to two-way by changing the switch by the door and adding a two-way pull cord switch over the bed.

As all conductors (except cpc's) between two-way switches are line conductors, it does not matter which colours go to which terminals, as long as the arrangement is the same at both switches and grey (old blue) and black (old yellow) are identified as brown (old red).

Two-way and intermediate switching

This is simply an extension of the two-way system, whereby one or more intermediate switches are wired between the two-way switches (Figure 2.15, see page 29).

Feeding the switch

As we have seen, the most common method of installing lighting circuits is a loop-in system, but there are occasions when this

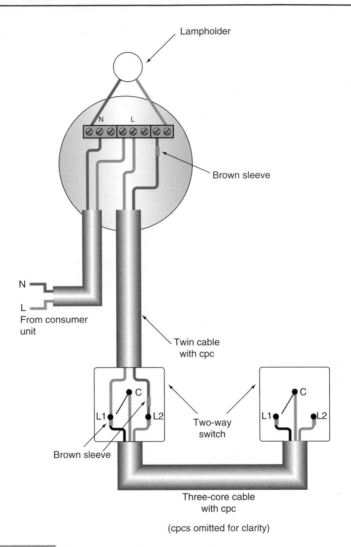

Lampholder

N L

Brown sleeve

N

L

From consumer unit

Twin cable with cpc

C

L1 L2

Two-way switch

Brown sleeve

C

L1 L2

Three-core cable with cpc

(cpcs omitted for clarity)

FIGURE 2.14 Two way switching via the switches.

system becomes difficult to wire satisfactorily. A typical example of this is the control of centre and wall lights in a large area. Consider the layout shown in Figure 2.16 and let us assume that ceiling points A and B are to be controlled by one switch and wall lights C, D and E are each controlled by a separate switch.

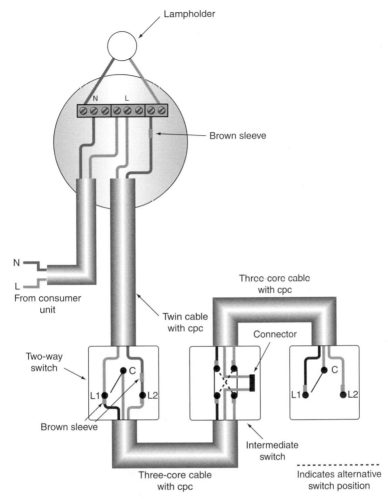

Lampholder

Brown sleeve

N

L

From consumer
unit

Three-core cable
with cpc

Twin cable
with cpc

Connector

Two-way
switch

C

L1   L2

Brown sleeve

Intermediate
switch

Three-core cable
with cpc

Indicates alternative
switch position

C

L1   L2

(cpcs omitted for clarity)

FIGURE 2.15   Two way and intermediate switching.

The symbols shown are BS EN 60617 architectural symbols. The × represents a lighting point, the ♂ indicates a one-way switch and the numeral 4 shows how many switches are required. In this case, the diagram shows four one-way switches, but it

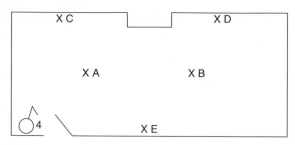

FIGURE 2.16   Light points controlled by feeding the switch.

FIGURE 2.17   Four gang switch.

would be normal to install a four-gang switch (Figure 2.17) rather than four separate one-way switches.

These multiple-gang switches are, however, manufactured as a bank of two-way switches and so not all the terminals are needed. Normally, the standard 'on-off' position is achieved by using the common terminal C and L1 (Figure 2.18).

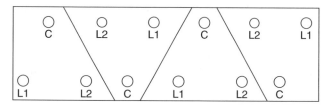

FIGURE 2.18   Terminal layout of a four gang switch.

Returning to the wiring of the lighting points shown in Figure 2.16, to use a loop-in system would be very difficult, as wall lights rarely come with enough terminals or space to accommodate three cables and the associated conductors. For example, a feed would enter at, say, point A and loop to points C, D and E. Then switch cables would need to be run back from each of these points to the four-gang switch: complicated, congested and expensive in cable.

Bringing a supply or feed to the four-gang switch and running single cables to each of the points to be controlled is far more sensible in all aspects (Figure 2.19).

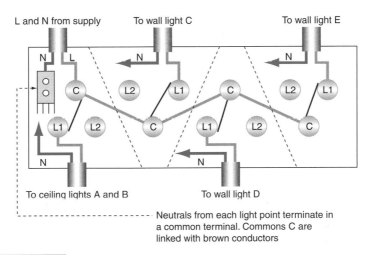

FIGURE 2.19 Wiring for feeding the switch.

Light point B (Figure 2.16) would, of course, be wired from the lampholder terminals of point A.

Neutrals from each light point terminate in a common terminal. Commons © are linked with brown (old red) conductors.

Radial socket-outlet circuits

Although most domestic installations use ring final circuits to supply socket outlets, radial circuits are quite acceptable. The recommendations for such circuits are given in Table 2.1. These radial circuits may have both fused and non-fused spurs (Figure 2.20).

Table 2.1 Coventional Circuit Arrangements for Radial Socket outlet Circuits.

Protective Device Size	Protective Device Type	Maximum Floor Area Served	Cable Size	Number of Socket Outlets
30 A or 32 A	any	75 m^2	4.0 m^2	unlimited
20 A	any	50 m^2	2.5 m^2	unlimited

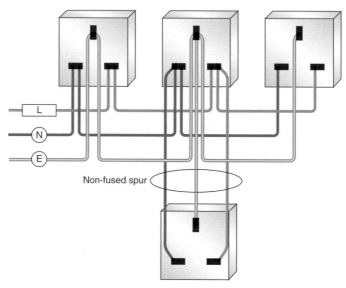

Non-fused spur

FIGURE 2.20 Non-fused spur.

Storage heater circuits

These, as discussed, are arranged to be energized during off-peak periods and it is usual to have a separate radial circuit for each heater, terminating in a fused connection unit.

Water heater circuits

Most small, over- or under-sink water heaters (less than 15 litres), storage or instantaneous, are fed from fused spurs from ring circuits. However, heaters over 15 l capacity and showers must be fed from their own individual radial circuits.

Cooker circuits

These radial circuits feed cooker units or control switches from which the cooking appliances are supplied. Cooker units are available with or without socket outlets.

If the rating of the circuit is between 15 and 50 A, one radial may feed two or more cooking appliances in the same room. Typical of this is the split-level cooker arrangement (i.e. an oven and a remote hob). The control switch must be within 2 m of any appliance and, in the case of two appliances, within 2 m of either one (Figure 2.21).

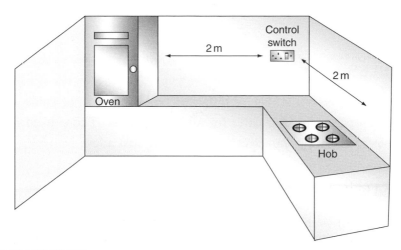

FIGURE 2.21   Position of cooker control switch.

Ring final circuits

Generally referred to as ring mains, these are the most common method of supplying BS 1363 socket outlets in domestic installations. The general requirements for these conventional circuits are as follows.

1. Provided that the ring does not serve an area exceeding $100\,m^2$, an unlimited number of sockets may be fed using either $2.5\,m^2$ PVC-insulated copper conductors or $1.5\,m^2$ mineral-insulated copper conductors (MICCs) and protected by a 30 or 32 A device of any type.
2. The number of **fused** spurs is unlimited.
3. There may be as many **non-fused** spurs as there are points on the ring.
4. Each non-fused spur may feed one single-socket outlet, one double-socket outlet or one item of permanently connected equipment.

Figure 2.22 illustrates these points.

CABLES

There are a vast number of different types and size of cable and conductors available to cater for all applications. Those used in a domestic situation, however, are limited to just a few.

Fixed wiring

Fixed wiring is the wiring that

- supplies all the outlets in the installation, sockets, lighting point, etc.
- connects together detectors, sensors, etc., for security systems
- supplies to telephone points and call systems.

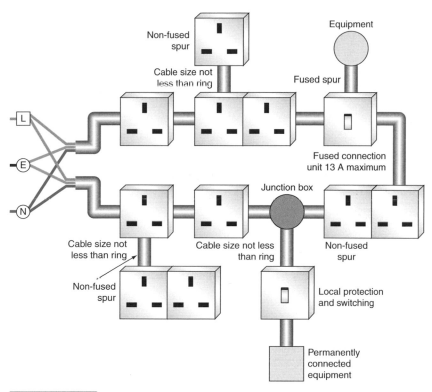

Non-fused
spur

Cable size not
less than ring

Equipment

Fused spur

L

E

N

Fused connection
unit 13 A maximum

Junction box

Cable size not
less than ring

Non-fused
spur

Cable size not less
than ring

Non-fused
spur

Local protection
and switching

Permanently
connected
equipment

FIGURE 2.22   Ring final circuit.

Power and lighting

The cable used for these applications will predominately be of
the flat twin with cpc variety (6242 Y) or flat three-core with cpc
(6243 Y) (Figure 2.23).

The use of steel or rigid PVC conduit containing single-core conductors is rare in domestic premises, with the exception of flats or
apartments with solid floors and ceilings and some older properties.

As we have seen, modern 6242 Y cable has brown and blue insulated conductors, the blue needing to be sleeved brown when used
as a switch wire in lighting circuits. However, a 6242 Y cable is

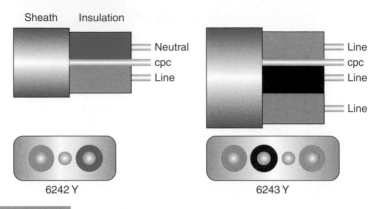

FIGURE 2.23 Flat profile cable.

available with both conductors coloured brown. Of course, when used for switch drops, this obviates the need for brown sleeving.

When underground supplies are required to supply remote garages, sheds, workshops and garden lighting, 6242 Y cable can be used, provided it is protected by running in galvanized conduit or is otherwise protected from mechanical damage. It is, however, better to use a steel-wire armoured (SWA) cable (Figure 2.24), as this type is specifically designed for harsh applications.

Alternatively, PVC-sheathed MICCs can be used (Figure 2.25), although this is expensive and requires special terminating tools.

The only other cables involved in the fixed wiring would be the single-core line, neutral and earth conductors at the main intake (tails) and bonding conductors. All are categorized as 6491 X, the line and neutral are PVC-sheathed (usually grey) as well as insulated. The earth conductors, collectively called protective conductors, are green and yellow.

Security

In this case, the cable used to link sensors and detectors will depend on the system, but is usually multicore PVC-sheathed

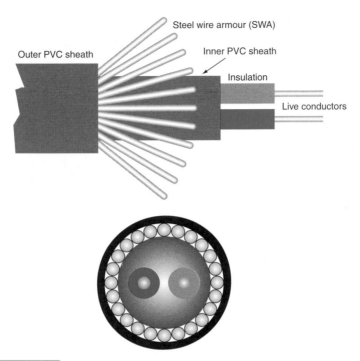

FIGURE 2.24 Steel wire armoured cable (SWA).

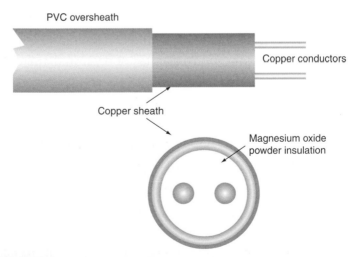

FIGURE 2.25 Mineral insulated cable.

and insulated with stranded conductors typically 7/0.2 (i.e. seven strands of 0.2 mm diameter wire).

Fire alarms

An increasing number of homes are now having fire alarm systems installed, very often linked to a security system. The cable used to link the sensors is usually FP 200 or Firetuff (both of which are fire retardant).

Telephone and call systems

These employ similar cable to that used for security systems.

Flexible cords

Flexible cords are flexible cables up to 4.0 mm^2 and are used to make the final connection from the fixed wiring to accessories or equipment either directly or via plugs. The conductors of these cords are made of many strands of thin wire, giving the flexible quality with which we are familiar. The number and size of these strands depend on the overall size of the conductor, for example a 0.5 mm^2 conductor comprises 16 strands of 0.2 mm diameter wire, whereas a 2.5 mm^2 conductor has 50 strands of 0.25 mm diameter wire.

There are a variety of flexible cords used in a domestic situation, varying from 0.5 to 2.5 mm^2 (4.0 mm^2 would hardly ever be used). The choice of cord type and size will depend on the appliance or equipment and the environmental conditions. So, for example, pendant drops from ceiling roses to lampholders would be wired with circular general-purpose PVC cords, whereas heat-resistant PVC or butyl rubber flex would be used for connection to heating appliances.

Note

Flexible cords should not be used for fixed wiring.

WIRING SYSTEMS AND INSTALLATION METHODS

There are two basic wiring systems used in domestic premises:

1. Flat twin and three-core cable (6242 Y and 6243 Y), clipped to the building structure, run in the floor, wall and ceiling voids and, for added protection when wiring is run surface, enclosed in mini-trunking.
2. Single-core cables (6491 X), enclosed in steel or rigid PVC conduit or in skirting and/or architrave trunking.

The flat sheathed cables are more likely to be used with the mini-trunking systems, which are not usually continuous throughout the whole installation.

Included in the trunking systems is the pre-wired flexible conduit system, which comprises a flat profile, flexible, compartmentalized conduit with PVC singles included (Figure 2.26). Conduit systems,

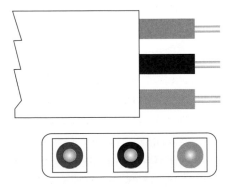

FIGURE 2.26   Pre-wired flexible conduit system.

pre-wired or otherwise, are usually installed where there are no building voids such as flats with solid floors and ceilings.

Installing flat twin and three-core cables

Before we begin to examine the methods of installing flat profile PVC cables, it is important to be aware of certain constraints.

When a cable is bent, the conductor insulation on the inside of the bend will be compressed and that on the outside stretched. Consequently, cable bends should be such that no damage to insulation is caused. For flat profile cables the minimum bending radius is three times the cable width for cables up to $10\,mm^2$ and four times for $16\,mm^2$ (Figure 2.27 and Table 2.2).

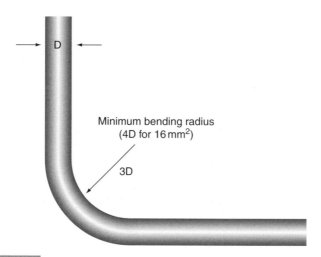

D

Minimum bending radius
($4D$ for $16\,mm^2$)

$3D$

FIGURE 2.27   Minimum bending radius.

Cables should be run without twists, and so should be removed from the cable drum in the same way as they were wound on. They should not be uncoiled, which would, of course, create

Table 2.2 Cable Bending Radii.

6242 Y			6243 Y		
Cable Size	Cable Width	Bend Radius	Cable Size	Cable Width	Bend Radius
$1.0\,mm^2$	7.6 mm	22.8 mm	$1.0\,mm^2$	9.9 mm	29.7 mm
$1.5\,mm^2$	8.6 mm	25.8 mm	$1.5\,mm^2$	11.4 mm	34.2 mm
$2.5\,mm^2$	10.2 mm	30.6 mm			
$4.0\,mm^2$	11.8 mm	35.4 mm			
$6.0\,mm^2$	13.4 mm	40.2 mm			
$10.0\,mm^2$	17.2 mm	51.6 mm			
$16.0\,mm^2$	19.4 mm	77.6 mm			

twists. Flat profile cables are designed to be clipped flat **not** on edge, as this may lead to damage by compressing the live conductor's insulation on to the bare cpc (Figure 2.28).

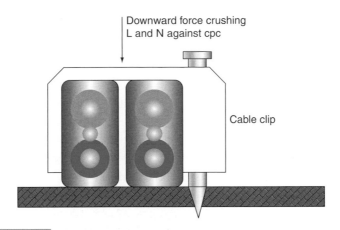

Downward force crushing L and N against cpc

Cable clip

FIGURE 2.28 Incorrect cable clipping.

Problems of embrittlement and cracking of PVC-sheathing and insulation can occur if PVC cable comes into contact with polystyrene. A reaction takes place between the two materials such that the plasticizer used in the manufacture of the PVC covering will migrate onto the polystyrene, leaving the cable in a rigid condition. If the cable is disturbed at a later date, the PVC is likely to crack, with the obvious consequences. Typical of this situation is the use of polystyrene granules, generally in older properties, as thermal insulation.

With these constraints in mind, let us now consider installation methods and techniques.

The IEE Wiring Regulations (BS 7671) give details of various installation methods. Those appropriate to domestic installations are known as methods A, B, C, 100, 102, 102, and 103 (Figures 2.29, 2.30 (page 44) and 2.31 (a), (b), (c) and (d) (page 45) illustrate these methods).

The first-fix

This is the term used to describe the initial installation of cables and associated accessories, prior to ceilings being boarded, walls being plastered and floors being laid. Before cables are drawn in it is, perhaps, best to install all of the metal knock-out (KO) boxes.

Those used for lighting circuits can usually be fixed directly to brick or block surfaces without 'chasing in', as they are designed to be plaster depth. All other boxes must be chased in such that their front edge will be flush with the finished plaster line (Figure 2.32, see page 46).

At first-fix stage, it is common practice to hold these metal boxes in place with galvanized plasterboard nails rather than screws and wall plugs, as the rendering and final plaster will prevent the box ever moving.

Where cables enter KO boxes, holes should have grommets installed. In general, it is usual to either clip cables direct to a surface and

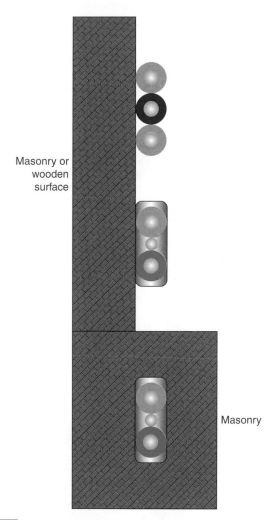

Masonry or wooden surface

Masonry

FIGURE 2.29 Method C.

give added protection from danger by the plasterer's trowel by covering the cable in PVC or metal Top-Hat section or enclose it in oval conduit.

There are, however, occasions when a specification demands that all cable runs are chased in to achieve full plaster cover. Also

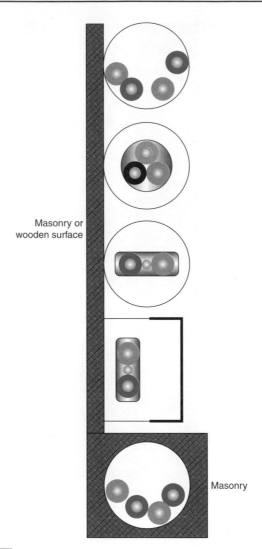

Masonry or
wooden surface

Masonry

FIGURE 2.30   Method B.

at first-fix stage and before cables are run, it is usual to drill
or notch joints. Great care must be taken when carrying out
such work as the strength of joints may be impaired by indiscrimi-
nate drillings or notching. The UK Building Regulations indicate

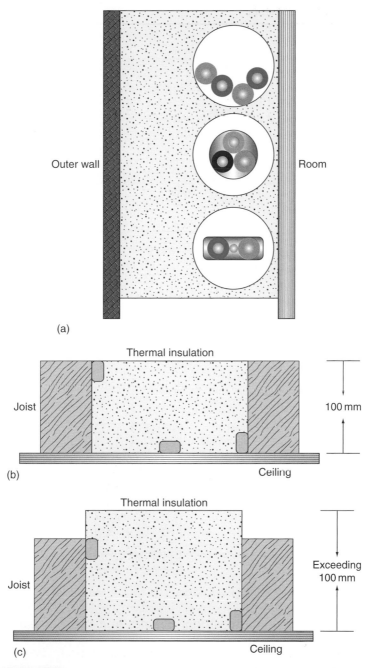

Outer wall

Room

(a)

Thermal insulation

Joist

100 mm

Ceiling

(b)

Thermal insulation

Joist

Exceeding
100 mm

Ceiling

(c)

FIGURE 2.31 (a) Method A, (b) Method 100, (c) Method 101, (d) Method 102, (e) Method 103.

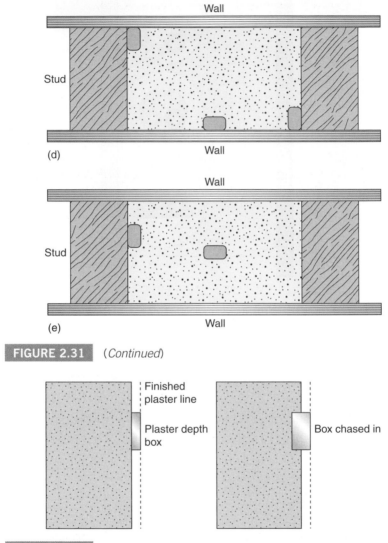

FIGURE 2.31 (*Continued*)

FIGURE 2.32 Fixing knock-out boxes.

the maximum size and position of holes and notches (shown in Figure 2.33).

Having carried out all the preliminary chasing, drilling and notching, cables can now be run in. Those that are installed in walls

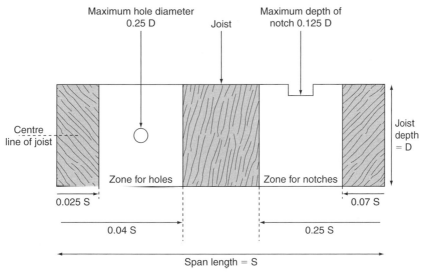

Maximum hole diameter 0.25 D
Joist
Maximum depth of notch 0.125 D

Centre line of joist

Joist depth = D

Zone for holes

Zone for notches

0.025 S

0.07 S

0.04 S

0.25 S

Span length = S

Holes for sheathed cables must be at least 50 mm from top or bottom of the joist

FIGURE 2.33 Holes and notches in joists.

or in partitions at a depth less than 50 mm should be run direct to the accessory either vertically or horizontally. They can be mechanically protected by metal conduit (not capping as this is too flimsy) to avoid penetration by nail, screws, etc., but where this is not practicable and such cable installation is in a premises used by unsupervised persons (typically domestic), the cables must be protected by a 30 mA RCD.

In order to avoid undue strain on conductors, cables in accessible positions must be supported at set intervals along both horizontal and vertical runs. These support intervals are shown in Table 2.3.

Clearly, these distances are not applicable to sheathed cables, enclosed or capped, if they are embedded in the building material, as they would be supported throughout their entire length.

Table 2.3 Supports by Clips.

Cable Type and Size (mm²)		Horizontal Support (mm)	Vertical Support (mm)
6242 Y	1.0	250	400
	1.5	250	400
	2.5	300	400
	4.0	300	400
	6.0	300	400
	10.0	350	450
	16.0	350	450
6243 Y	1.0	300	400
	1.5	300	400

When cables are installed in floor voids in notches in joists, they must be protected by an earth metal covering (e.g. conduit).

Where cables are run to a lighting point on a plasterboard ceiling and a joist is conveniently placed, cables are clipped to the side and the ceiling rose or batten holder secured to the underside of the joist (Figure 2.34).

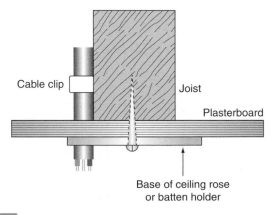

FIGURE 2.34   Fixing to joist.

However, if the location of the point is in between joists, a 'noggin' is fixed between to enable the fitting to be fixed securely (Figure 2.35).

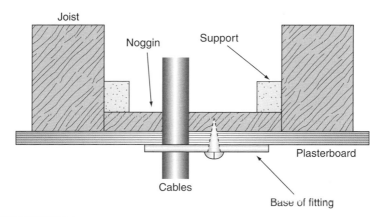

FIGURE 2.35 Fixing to noggin.

The same sort of fixing arrangement applies to KO boxes in stud walling (Figure 2.36).

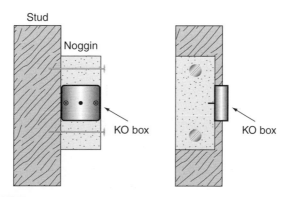

FIGURE 2.36 Fixing KO box to stud wall.

Conduit systems

Due to expense, both of material and labour, it is not a common practice to install a conduit wiring system, metal or PVC, in domestic situations. However, where such systems are specified, the whole conduit system must be erected before any cables are drawn in.

Clearly, with such systems, all conduit runs will need to be 'chased in' and joists notched. Supporting distances for surface conduits, if used in domestic situations, are shown in Table 2.4.

Table 2.4 Supports for Conduits.

	Maximum Distance Between Supports			
	Metal		**PVC**	
Conduit diameter	Horizontal	Vertical	Horizontal	Vertical
20–25 mm	1.75 m	2.0 m	1.5 m	1.75 m

For a 'first-fix' on a concealed conduit system, crampets would hold the conduit in place until the render and plaster were applied (Figure 2.37).

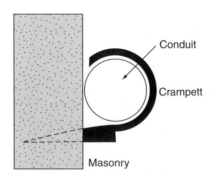

Conduit

Crampett

Masonry

FIGURE 2.37 Crampett fixing.

Conduit systems should be complete and continuous through-out and so entries to KO boxes will be via male or female bushes (Figure 2.38).

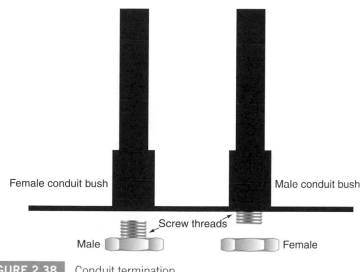

Female conduit bush

Male conduit bush

Screw threads

Male

Female

FIGURE 2.38 Conduit termination.

Whatever system is used, it is sensible to leave plenty of cable/con-ductor at each outlet at first-fix stage (150 mm is reasonable).

The second-fix

Fitting accessories is the nice, clean part! It is at this point that some degree of testing could be carried out to determine if any damage to cables has occurred since the first-fix was completed (see Chapter 6). Clearly, in order to fit accessories, cable sheaths and/or insulation must be removed. This is the subject of terminations.

Terminations

Great care must be taken when terminating cables and conductors in order to avoid accidental damage from terminating tools.

Preparing flat profile cables

In order to terminate flat profile cables, the outer sheath must be removed. There are two ways of achieving this, using a knife or using the cable cpc:

1. The knife is used to slice the sheathing lengthways and then to cut off the surplus (Figure 2.39).

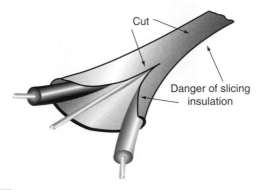

FIGURE 2.39 Terminating flat cable.

2. The cable end is split with a pair of cutters and the cpc exposed and hooked out. This may be held by pliers and dragged backwards along the cable, splitting the sheath. The surplus is cut off as in Figure 2.40.

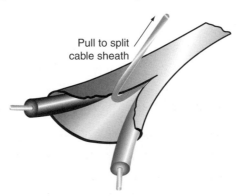

FIGURE 2.40 Terminating flat cable.

Sheathing should be removed close to the entry to an accessory, but must not be removed outside the accessory enclosure. Too much sheath left on a cable inside an enclosure makes the whole termination process difficult as space is at a premium (Figure 2.41).

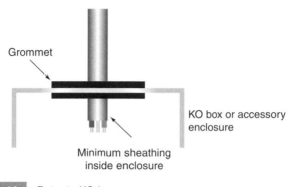

Grommet

KO box or accessory enclosure

Minimum sheathing inside enclosure

FIGURE 2.41 Entry to KO box.

With flexible cords, the process of sheath removal is probably best using a knife. Here, the sheathing is very carefully cut around the cable at the desired length, until the colours of the cores are just visible. The sheath can then usually be pulled off. Care must be taken to avoid any cutting of the conductor insulation.

Preparing mineral-insulated cables

Although mineral-insulated cable installations are uncommon in domestic installations, it is worth taking a brief look at the termination of such cable. Special tools are required to strip the outer sheath and to assemble all the respective parts of the termination.

First, the outer sheath is removed to the required length (care being taken to avoid any moisture being absorbed by the magnesium oxide powder insulation). Then, the various parts of the termination assembly are slid onto the cable, compressed and screwed together (Figure 2.42).

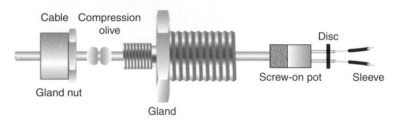

FIGURE 2.42   Mineral cable termination.

Preparing steel wire armoured cable

In a domestic situation, probably the only time this cable would be used would be for supplies to garages, workshops, garden lighting, etc. Figure 2.43 shows the component parts of an SWA cable termination.

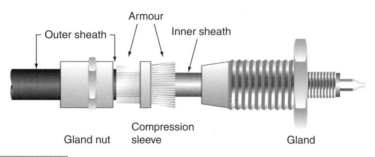

FIGURE 2.43   SWA termination.

Once sheathing has been removed from flat profile cables, flexible cords and mineral-insulated or SWA cables have been prepared, the next stage is to strip the insulation from the conductors. This must also be done with the utmost care. Conductors should not be scored nor have strands removed. Only the minimum amount of insulation should be removed to enable the conductor to be housed in its terminal (Figure 2.44).

When single small conductors are to be housed in large terminals, it is usual to double over the conductor to ensure a good contact with the terminal screw (Figure 2.45).

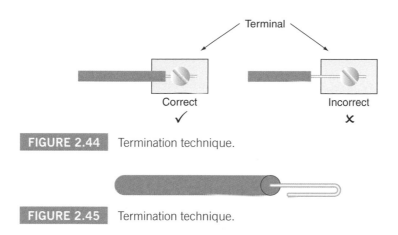

Terminal

Correct
✓

Incorrect
✗

FIGURE 2.44 Termination technique.

FIGURE 2.45 Termination technique.

Do not twist conductors together when terminating, it strains them and makes disconnection and reconnection before and after testing very difficult. Always leave sufficient conductor to enable easy entry to accessory terminals, never cut them to exact length, as it will be difficult to effect an efficient termination.

CHAPTER 3

The Importance of Earthing and Bonding

ELECTRIC SHOCK

This is the passage of current through the body of such magnitude as to have significant harmful effects. Table 3.1 and Figure 3.1 illustrate the generally accepted effects of current passing through the human body. How, then, are we at risk of electric shock and how do we protect against it?

There are two ways in which we can be at risk:

1. Touching live parts of equipment or systems that are intended to be live.
2. Touching conductive parts which are not meant to be live, but which have become live due to a fault.

The conductive parts associated with condition (2) can either be metalwork of electrical equipment and accessories (Class I) and that

Table 3.1 Effects of Current Passing Through the Human Body.

1–2 mA	Barely perceptible, no harmful effects
5–10 mA	Throw off, painful sensation
10–15 mA	Muscular contraction, can't let go
20–30 mA	Impaired breathing
50 mA and above	Ventricular fibrillation and death

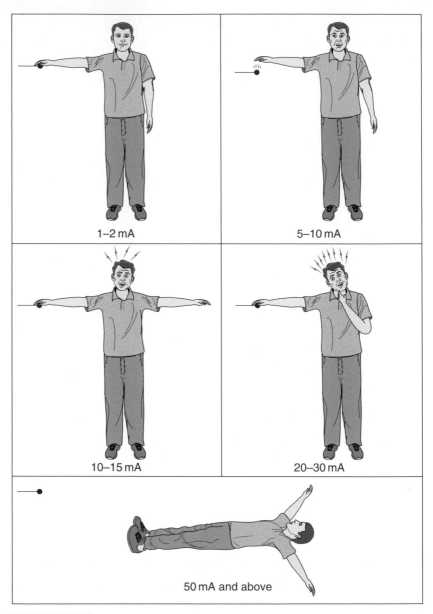

1–2 mA

5–10 mA

10–15 mA

20–30 mA

50 mA and above

FIGURE 3.1 Shock levels.

of electrical wiring systems (e.g. metal conduit and trunking), called **exposed conductive** parts, or other metalwork (e.g. pipes, radiators and girders), called **extraneous conductive** parts.

BASIC PROTECTION

How can we prevent danger to persons and livestock from contact with intentionally live parts? Clearly, we must minimize the risk of such contact and this can be achieved by basic protection which is

- insulating any live parts or
- ensuring any uninsulated live parts are housed in suitable enclosures and/or are behind barriers.

The use of a residual current device (RCD) cannot prevent this contact, but it can be used as additional protection to any of the other measures taken, provided that its rating, $I_{\Delta n}$, is 30 mA or less and has an operating time of not more than 40 ms at a test current of $I_{\Delta n} \times 5$.

It should be noted that RCDs are not the panacea for all electrical ills, they can malfunction, but they are a valid and effective back-up to the other methods. They must not be used as the sole means of protection.

FAULT PROTECTION

How can we protect against shock from contact with unintentionally live, exposed or extraneous conductive parts whilst touching earth, or from contact between live exposed and/or extraneous conductive parts? The most common method is by protective earthing, protective equipotential bonding and automatic disconnection in case of a fault.

All extraneous conductive parts are joined together by main protective bonding conductors and connected to the main earthing terminal, and all exposed conductive parts are connected to the main earthing terminal by the circuit protective conductors (cpc's). Add to this overcurrent protective devices that will operate fast enough when a fault occurs and the risk of severe electric shock is significantly reduced.

WHAT IS EARTH AND WHY AND HOW DO WE CONNECT TO IT?

The thin layer of material which covers our planet – rock, clay, chalk or whatever – is what we in the world of electricity refer to as earth. So, why do we need to connect anything to it? After all, it is not as if earth is a good conductor.

It might be wise at this stage to investigate potential difference (PD). A PD is exactly what it says it is: a difference in potential (volts). In this way, two conductors having PDs of, say, 20 and 26 V have a PD between them of $26 - 20 = 6$ V. The original PDs (i.e. 20 and 26 V) are the PDs between 20 V and 0 V and 26 V and 0 V. So where does this 0 V or zero potential come from? The simple answer is, in our case, the earth. The definition of earth is, therefore, the conductive mass of earth, whose electric potential at any point is conventionally taken as zero.

Thus, if we connect a voltmeter between a live part (e.g. the line conductor of a socket outlet) and earth, we may read 230 V; the conductor is at 230 V and the earth at zero. The earth provides a path to complete the circuit. We would measure nothing at all if we connected our voltmeter between, say, the positive 12 V terminal of a car battery and earth, as in this case the earth plays no part in any circuit.

Figure 3.2 illustrates this difference.

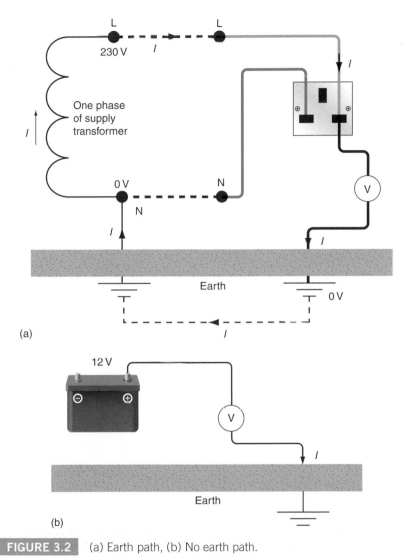

(a)

(b)

FIGURE 3.2 (a) Earth path, (b) No earth path.

So, a person in an installation touching a live part whilst stand-
ing on the earth would take the place of the voltmeter and could
suffer a severe electric shock. Remember that the accepted lethal

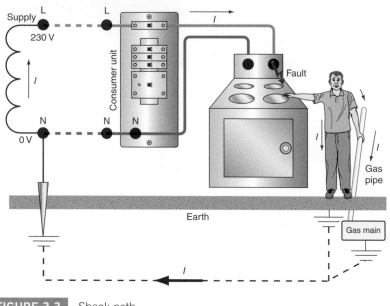

FIGURE 3.3   Shock path.

level of shock current passing through a person is only 50 mA or 1/20 A. The same situation would arise if the person were touching a faulty appliance and a gas or water pipe (Figure 3.3).

One method of providing some measure of protection against these effects is, as we have seen, to join together (bond) all metallic parts and connect them to earth. This ensures that all metalwork in a healthy installation is at or near 0 V and, under fault conditions, all metalwork will rise to a similar potential. So, simultaneous contact with two such metal parts would not result in a dangerous shock, as there would be no significant PD between them.

Unfortunately, as mentioned, earth itself is not a good conductor, unless it is very wet. Therefore, it presents a high resistance to the flow of fault current. This resistance is usually enough to restrict fault current to a level well below that of the rating of the protective device, leaving a faulty circuit uninterrupted. Clearly this is an unhealthy situation.

In all but the most rural areas, consumers can connect to a metallic earth return conductor, which is ultimately connected to the earthed neutral of the supply. This, of course, presents a low-resistance path for fault currents to operate the protection.

In summary, connecting metalwork to earth places that metal at or near zero potential and bonding between metallic parts puts such parts at a similar potential even under fault conditions. Add to this a low-resistance earth fault return path, which will enable the circuit protection to operate very fast, and we have significantly reduced the risk of electric shock.

Earth fault loop impedance

As we have just seen, circuit protection should operate in the event of a fault to earth. The speed of operation of the protective device is of extreme importance and will depend on the impedance of the earth fault loop path.

Figure 3.4 shows this path. Starting at the point of the fault, the path comprises

- the circuit protective conductor (cpc)
- the consumer's earthing terminal and earthing conductor
- the return path, either metallic or earth itself
- the earthed neutral of the supply transformer
- the transformer winding
- the line conductor from the transformer to the fault.

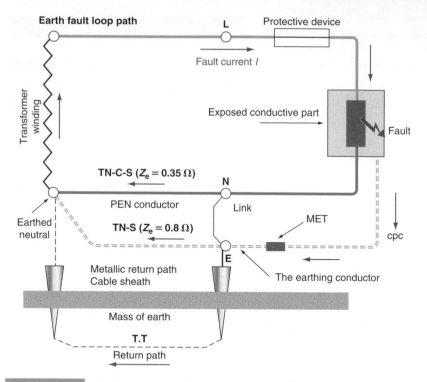

Earth fault loop path   L   Protective device

Fault current *I*

Transformer winding

Exposed conductive part

Fault

TN-C-S ($Z_e = 0.35\ \Omega$)   N

PEN conductor   Link

MET

Earthed neutral   **TN-S ($Z_e = 0.8\ \Omega$)**

cpc

E

Metallic return path
Cable sheath   The earthing conductor

Mass of earth

T.T
Return path

FIGURE 3.4   Earth fault loop path.

Figure 3.5 is a simplified version of the loop path.

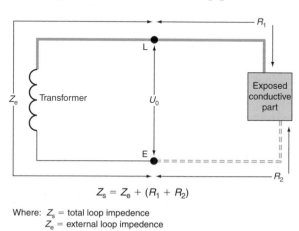

R_1

L

Exposed conductive part

Z_e Transformer U_0

E

R_2

$$Z_s = Z_e + (R_1 + R_2)$$

Where: Z_s = total loop impedence
Z_e = external loop impedence
R_1 = resistance of the circuit line conductor
R_2 = resistance of the circuit cpc

FIGURE 3.5   Simplified earth fault loop path.

From Figure 3.5, we can see that the total earth fault loop impedance (Z_s) is made up of the impedance external to the installation (Z_e), the resistance of the circuit line conductor (R_1) and that of the circuit cpc (R_2), i.e.

$$Z_s = Z_e + R_1 + R_2$$

we also have, from Ohm's law, the value of the fault current that would flow from

$$I = \frac{U_0}{Z_s}$$

where U_0 is the nominal voltage to earth (230 V).

Determining the value of total loop impedance Z_s

The IEE Regulations require that when the general characteristics of an installation are assessed, the loop impedance Z_e external to the installation shall be ascertained. This may be measured in existing installations using a line-to-earth loop impedance tester. However, when a building is only at the drawing-board stage, it is clearly impossible to make such a measurement. In this case, we have three methods available to assess the value of Z_e.

1. Determine it from details (if available) of the supply transformer, the main distribution cable and the proposed service cable.
2. Measure it from the supply intake position of an adjacent building which has service cable of similar size and length to that proposed.
3. Use the maximum likely values issued by the supply authority, as follows:
   (a) TT system: 21 Ω maximum
   (b) TN-S system: 0.8 Ω maximum
   (c) TN-C-S system: 0.35 Ω maximum.

Method 1 will be difficult for anyone except engineers. Method 3 can, in some cases, result in pessimistically large cable sizes. Method 2, if it is possible to use it, will give a closer and more realistic estimation of Z_e. However, if in any doubt, use Method 3.

Having established a value for Z_e, it is now necessary to determine the impedance of that part of the loop path internal to the installation. This is, as we have seen, the resistance of the line conductor plus the resistance of the cpc (i.e. $R_1 + R_2$). Resistances of copper conductors may be found from manufacturers' information, which gives values of resistance/metre for copper and aluminium conductors at 20°C in mΩ/m. Table 3.2 gives resistance values for copper conductors up to 35 mm^2.

So, a 25.0 mm^2 line conductor with a 4.0 mm^2 cpc has $R_1 = 0.727$ mΩ and $R_2 = 4.61$ mΩ, giving $R_1 + R_2 = 0.727 + 4.61 = 5.337$ mΩ/m. So, having established a value for $R_1 + R_2$, we must now multiply it by the length of the run and divide by 1000 (the values given are in milli-ohms per metre). However, this final

Table 3.2 Resistance of Copper Conductors in mΩ/m at 20°C.

Conductor (mm^2)	Resistance (mΩ)
1.0	18.1
1.5	12.1
2.5	7.41
4.0	4.61
6.0	3.08
10.0	1.83
16.0	1.15
25.0	0.727
35.0	0.524

value is based on a temperature of 20°C, but when the conductor is fully loaded, its temperature will increase. In order to determine the value of resistance at conductor-operating temperature, a multiplier is used. This multiplier, applied to the 20°C value of resistance, is 1.2 for PVC cables.

Hence, for a 20 m length of 70°C PVC-insulated 16.0 mm^2 line conductor with a 4.0 mm^2 cpc, the value of $R_1 + R_2$ would be

$$R_1 + R_2 = [(1.15 + 4.61) \times 20 \times 1.2] / 1000 = 0.138\,\Omega$$

We are now in a position to determine the total earth fault loop impedance Z_s from

$$Z_s = Z_e + R_1 + R_2$$

As mentioned, this value of Z_s should be as low as possible to allow enough fault current to flow to operate the protection as quickly as possible. The IEE Regulations give maximum values of loop impedance for different sizes and types of protection for final circuits not exceeding 32 A and distribution circuits. Provided that the actual calculated values do not exceed those tabulated, final circuits will disconnect under earth fault conditions in 0.4 s or less, and distribution circuits in 5 s or less. The reasoning behind these different times is based on the time that a faulty circuit can reasonably be left uninterrupted and is based on the probable chances of someone being in contact with exposed or extraneous conductive parts at the precise moment that a fault develops.

Example 2.1

Let's have a look at a typical example of a shower circuit run in an 18 m length of 6.0 mm^2 (6242 Y) twin cable with cpc, and protected by a 30 A BS 3036 semi-enclosed rewirable fuse. A 6.0 mm^2 twin cable has a 2.5 mm^2 cpc. We will also assume that the external loop impedance Z_e is measured at 0.27 Ω. Will there be a shock risk if a phase-to-earth fault occurs?

The total loop impedance $Z_s = Z_e + R_1 + R_2$ and we are given $Z_e = 0.27\,\Omega$. For a 6.0 mm^2 line conductor with a 2.5 mm^2 cpc, $R_1 + R_2$ is 10.49 mΩ/m. Hence, with a multiplier of 1.2 for 70°C PVC, the total $R_1 + R_2 = 18 \times 10.49 \times 1.2/1000 = 0.23\,\Omega$. Therefore, $Z_s = 0.27 + 0.23 = 0.53\,\Omega$. This is less than the 1.09 Ω maximum given in the IEE Regulations for a 30 A BS 3036 fuse. Consequently, the protection will disconnect the circuit in less than 0.4 s.

RESIDUAL CURRENT DEVICES

We have seen how very important the total earth loop impedance Z_s is in the reduction of shock risk. However, in TT systems where the mass of earth is part of the fault path, the maximum values of Z_s given in the IEE Regulations may be hard to satisfy. Added to this, climatic conditions will alter the resistance of the earth in such a way that Z_e may be satisfactory in wet weather, but not in very dry.

The regulations recommend, therefore, that the preferred method of earth fault protection for installations in TT systems be achieved by RCDs, such that the product of its residual operating current and the loop impedance will not exceed a figure of 50 V. Residual current breakers (RCBs), residual current circuit breakers (RCCBs) and RCDs are one and the same thing. Modern developments in CB (circuit breaker), RCD and consumer unit design now make it easy to protect any individual circuit with a combined CB/RCD (RCBO), making the use of split-load boards unnecessary.

In domestic premises the use of 30 mA RCDs is required for the protection of all socket outlets rated at not more than 20 A, for all circuits in a bath or shower room and for cables embedded in walls and partitions at a depth less than 50 mm. Socket outlets not intended for general use, for example those provided for non-portable equipment such as freezers, etc., are exempt from this requirement, provided they are suitably labelled or identified (Table 3.3).

Table 3.3

30 mA
- All socket outlets rated at not more than 20 A and for unsupervised general use.
- Mobile equipment rated at not more than 32 A for use outdoors.
- All circuits in a bath/shower room.
- Preferred for all circuits in a TT system.
- All cables installed less than 50 mm from the surface of a wall or partition (in the safe zones) if the installation is unsupervised, and also at any depth if the construction of the wall or partition includes metallic parts.
- In zones 0, 1 and 2 of swimming pool locations.
- All circuits in a location containing saunas, etc.
- Socket outlet final circuits not exceeding 32 A in agricultural locations.
- Circuits supplying Class II equipment in restrictive conductive locations.
- Each socket outlet in caravan parks and marinas and final circuit for houseboats.
- All socket outlet circuits rated not more than 32 A for show stands, etc.
- All socket-outlet circuits rated not more than 32 A for construction sites (where reduced low voltage, etc. is not used).
- All socket outlets supplying equipment outside mobile or transportable units.
- All circuits in caravans.
- All circuits in circuses, etc.
- A circuit supplying Class II heating equipment for floor and ceiling heating systems.

500 mA
- Any circuit supplying one or more socket outlets of rating exceeding 32 A, on a construction site.

300 mA
- At the origin of a temporary supply to circuses, etc.
- Where there is a risk of fire due to storage of combustible materials.
- All circuits (except socket outlets) in agricultural locations.

100 mA
- Socket outlets of rating exceeding 32 A in agricultural locations.

Where Loop Impedances are too high, RCD ratings can be calculated.

BONDING: QUESTIONS AND ANSWERS

By now, we should know why bonding is necessary. The next question is to what extent bonding should be carried out? This is, perhaps, answered best by means of question and answer examples.

1. **Do I need to bond the kitchen hot and cold taps and a metal sink together?**

   The IEE Regulations do not require this bonding to be carried out in domestic kitchens/utility rooms, etc.

2. **Do I have to bond radiators in a premises to, say, metal clad switches or socket outlets?**

   Supplementary bonding is only necessary when extraneous conductive parts are simultaneously accessible with exposed conductive parts and when the disconnection time for the circuit concerned cannot be achieved. In these circumstances, the bonding conductor should have a resistance $R < 50/I_a$, where R = resistance of supplementary bonding conductor, 50 = touch voltage of 50 V and I_a = current causing operation of protection.

3. **Do I need to bond metal window frames?**

   In general, no! Apart from the fact that most window frames will not introduce a potential from anywhere, the part of the window most likely to be touched is the opening portion, to which it would not be practicable to bond. There may be a case for the bonding of patio doors, which could be considered earthy with rain running from the lower portion to the earth. However, once again, the part most likely to be touched is the sliding section, to which it is not possible to bond. In any case, there would need to be another simultaneously accessible part to warrant considering any bonding.

4. **What about bonding in bathrooms?**

   This is dealt with in detail in Chapter 4.

4. **What size of bonding conductors should I use?**

   Main protective bonding conductors should be not less than half the size of the main earthing conductor, subject to a minimum of $6.0\,mm^2$ or, where PME (TN-C-S) conditions are present, $10.0\,mm^2$. For example, most new domestic installations now have a $16.0\,mm^2$ earthing conductor, so all main bonding will be in $10.0\,mm^2$. Supplementary bonding conductors are subject to a minimum of $2.5\,mm^2$ if mechanically protected or $4.0\,mm^2$ if not. However, if these bonding conductors are connected to exposed conductive parts, they must be the same size as the cpc connected to the exposed conductive part (once again subject to the minimum sizes mentioned). It is sometimes difficult to protect a bonding conductor mechanically throughout its length, especially at terminations, so it is perhaps better to use $4.0\,mm^2$ as the minimum size.

5. **Do I have to bond free-standing metal cabinets, screens, work benches, etc.?**

   No! These items will not introduce a potential into the equipotential zone from outside and cannot, therefore, be regarded as extraneous conductive parts.

6. **What do I do when the pipework is plastic or a mixture of metal and plastic?**

   No bonding is required.

THE FARADAY CAGE

In one of his many experiments, Michael Faraday (1791–1867) placed himself in an open-sided cube, which was then covered in a conducting material and insulated from the floor. When this cage arrangement was charged to a high voltage, he found that he could move freely within it touching any of the sides, with no adverse effects. He had, in fact, created an equipotential zone. Of course, in a correctly bonded installation, we live and/or work in Faraday cages!

CHAPTER 4

Bathrooms

This chapter deals with common locations containing baths, showers and cabinets containing a shower and/or bath. It does not apply to specialist locations. The main feature of this section is the division of the location into zones (0, 1 and 2).

ZONE 0

This is the interior of the bath tub or shower basin or, in the case of a shower area without a tray, it is the space having a depth of 100 mm above the floor out to a radius of 600 mm from a fixed shower head or 1200 mm radius for a demountable head (Figure 4.1).

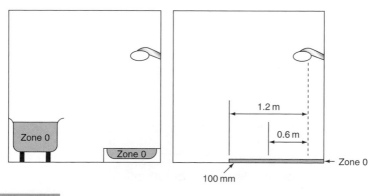

FIGURE 4.1   Zone 0.

- Only SELV (12V) or ripple-free DC may be used as a measure against electric shock, the safety source being outside zones 0, 1 and 2.
- Other than current using equipment specifically designed for use in this zone, **no** switchgear or accessories are permitted.
- Equipment designed for use in this zone must be to at least IPX7.
- Only wiring associated with equipment in this zone may be installed.

ZONE 1

This extends above zone 0 around the perimeter of the bath or shower basin to 2.25 m above the floor level, and includes any space below the bath or basin that is accessible without the use of a key or tool. For showers without basins, zone 1 extends out to a radius of 600 mm from a fixed shower head (Figure 4.2).

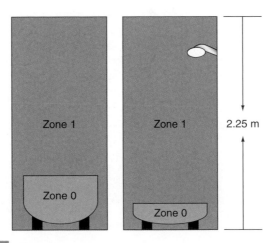

FIGURE 4.2   Zone 1.

ZONE 2

This extends 600 mm beyond zone 1 and to a height of 2.25 m above floor level (Figure 4.3).

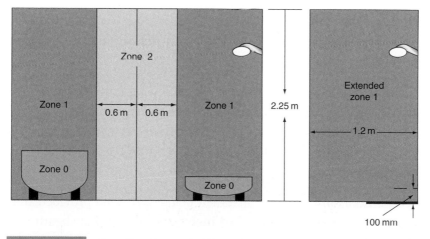

FIGURE 4.3 Zone 2 and extended Zone 1.

Note

■ Other than switches and controls of equipment specifically designed for use in this zone, and cord operated switches, only SELV switches are permitted.

■ Equipment designed for use in this zone must be to at least IPX4, or IPX5 where water jets are likely to be used for cleaning purposes.

■ For showers without basins there is no zone 2, just an extended zone 1.

■ Socket outlets other than SELV may **not** be installed within 3 m of the boundary of zone 1.

SUPPLEMENTARY EQUIPOTENTIAL BONDING

Supplementary bonding may be established connecting together the cpc's, exposed and extraneous conductive parts within the location.

Such extraneous conductive parts will include:

■ metallic gas, water, waste and central heating pipes
■ metallic structural parts that are accessible to touch
■ metal baths and shower basins.

This bonding may be carried out inside or outside the location preferably close to the entry of the extraneous conductive parts to the location.

However, this bonding may be omitted if the premises has a protective earthing and automatic disconnection system in place, all extraneous conductive parts of the locations are connected to the protective bonding and all circuits are RCD protected (which they have to be anyway!!).

Electric floor units may be installed below any zone provided that they are covered with an earthed metal grid or metallic sheath and connected to the protective conductor of the supply circuit.

CHAPTER 5

Protection

The meaning of the word 'protection', as used in the electrical industry, is no different to that in everyday use. People protect themselves against personal or financial loss by means of insurance and from injury or discomfort by the use of the correct protective clothing. They further protect their property by the installation of security measures such as locks and/or alarm systems.

In the same way, electrical systems need

- to be protected against mechanical damage, the effects of the environment, and electrical overcurrents
- to be installed in such a fashion that persons and/or livestock are protected from the dangers that such an electrical installation may create.

Let us now look at these protective measures in more detail.

PROTECTION AGAINST MECHANICAL DAMAGE

The word 'mechanical' is somewhat misleading, in that most of us associate it with machinery of some sort. In fact, a serious electrical overcurrent, left uninterrupted for too long, can cause distortion of conductors and degradation of insulation. Both of these effects are considered to be mechanical damage.

However, let us start by considering the ways of preventing mechanical damage caused by physical impact and the like.

Cable construction

As we have seen in Chapter 2, a cable comprises one or more conductors, each covered with an insulating material. This insulation provides protection from shock by bodily contact with live parts and prevents the passage of leakage currents between conductors. Clearly, insulation is very important and, in itself, should be protected from damage. This may be achieved by covering the insulated conductors with a protective sheathing during manufacture or by enclosing them in conduit or trunking at the installation stage.

The type of sheathing chosen and/or the installation method will depend on the environment in which the cable is to be installed. For example, in an environment subject to mechanical damage, metal conduit with polyvinyl chloride (PVC) singles or mineral-insulated cable would be used in preference to PVC-sheathed cable clipped direct.

Protection against corrosion

Mechanical damage to cable sheaths and the metalwork of wiring systems can occur through corrosion, and so care must be taken to choose corrosion-resistant materials and to avoid contact between dissimilar metals in damp situations.

Protection against thermal effects

The IEE Regulations basically require commonsense decisions regarding the placing of fixed equipment, so that surrounding materials are not at risk from damage by heat. Added to these requirements is the need to protect persons and livestock from burns by guarding parts of equipment liable to excessive temperatures.

Polyvinyl chloride

PVC is a thermoplastic polymer widely used in electrical installation work for cable insulation, conduit and trunking. General-purpose PVC is manufactured to BS 6746.

PVC in its raw state is a white powder; it is only after the addition of plasticizers and stabilizers that it acquires the form with which we are familiar.

Degradation

All PVC polymers are degraded or reduced in quality by heat and light. Special stabilizers added during manufacture help to retard this degradation at high temperatures. However, it is recommended that PVC-sheathed cables or thermoplastic fittings for luminaires (light fittings) should not be installed where the temperature is likely to rise above 60°C. Cables insulated with high-temperature PVC (up to 80°C) should be used for drops to lampholders and entries into batten holders. PVC conduit and trunking should not be used in temperatures above 60°C.

Embrittlement and cracking

PVC exposed to low temperatures becomes brittle and will crack easily if stressed. Although both rigid and flexible PVC used in cables and conduit can reach as low as −5°C without becoming brittle, the regulations recommend that general-purpose PVC-insulated cables should not be installed in areas where the temperature is likely to be consistently below 0°C. It is further recommended that PVC-insulated cable should not be handled unless the ambient temperature is above 0°C and unless the cable temperature has been above 0°C for at least 24 hours.

Where rigid PVC conduit is to be installed in areas where the ambient temperature is below $-5°C$ but not lower than $-25°C$, type B conduit manufactured to BS 4607 should be used.

When PVC-insulated cables are installed in loft spaces insulated with polystyrene granules, contact between the two polymers can cause the plasticizer in the PVC to migrate to the granules. This causes the PVC to harden and, although there is no change in the electrical properties, the insulation may crack if disturbed.

Protection against ingress of solid objects and liquids

In order to protect live parts of equipment being contacted by foreign solid bodies or liquid, and also to prevent persons or livestock from coming into contact with live or moving parts, such equipment is housed inside an enclosure.

The degree of protection offered by such an enclosure is indicated by an index of protection (IP) code, as shown in Table 5.1.

It will be seen from Table 5.1 that, for instance, an enclosure to IP 56 is dustproof and waterproof.

So, typical IP codes that would be experienced in the domestic installation would be IPXXB or IP2X and IPXXD or IP4X (IPXXB is protection against finger contact only and IPXXD protection against penetration by 1 mm diameter wires).

PROTECTION OF PERSONS/LIVESTOCK AGAINST DANGERS

Protection against electric shock

There are two ways of receiving an electric shock, both of which have been discussed in depth in Chapter 3.

Table 5.1   IP Codes.

First numeral: Mechanical protection

0. No protection of persons against contact with live or moving parts inside the enclosure. No protection of equipment against ingress of solid foreign bodies.
1. Protection against accidental or inadvertent contact with live or moving parts inside the enclosure by a large surface of the human body (e.g. a hand) but not protection against deliberate access to such parts. Protection against ingress of large solid foreign bodies.
2. Protection against contact with live or moving parts inside the enclosure by fingers. Protection against ingress of medium-size solid foreign bodies (12.5 mm spheres).
3. Protection against contact with live or moving parts inside the enclosure by tools, wires or such objects of thickness greater than 2.5 mm. Protection against ingress of small foreign bodies.
4. Protection against contact with live or moving parts inside the enclosure by tools, wires or such objects of thickness greater than 1 mm. Protection against ingress of small solid foreign bodies.
5. Complete protection against contact with live or moving parts inside the enclosure. Protection against harmful deposits of dust. The ingress of dust is not totally prevented, but dust cannot enter in an amount sufficient to interfere with satisfactory operation of the equipment enclosed.
6. Complete protection against contact with live or moving parts inside the enclosures. Protection against ingress of dust.

Second numeral: Liquid protection

0. No protection.
1. Protection against drops of condensed water. Drops of condensed water falling on the enclosure shall have no harmful effect.
2. Protection against drops of liquid. Drops of falling liquid shall have no harmful effect when the enclosure is tilted at any angle up to 15° from the vertical.
3. Protection against rain. Water falling in rain at an angle equal to or smaller than 60° with respect to the vertical shall have no harmful effect.
4. Protection against splashing. Liquid splashed from any direction shall have no harmful effect.
5. Protection against water jets. Water projected by a nozzle from any direction under stated conditions shall have no harmful effect.
6. Protection against conditions on ships' decks (deck with water-tight equipment). Water from heavy seas shall not enter the enclosures under prescribed conditions.
7. Protection against immersion in water. It must not be possible for water to enter the enclosure under stated conditions of pressure and time.
8. Protection against indefinite immersion in water under specified pressure. It must not be possible for water to enter the enclosure.

X   Indicates no specified protection.

Protection against overcurrent

An overcurrent is a current greater than the rated current of a circuit. It may occur in two ways:

1. As an overload current.
2. As a short circuit or fault current.

These conditions need to be protected against in order to avoid damage to circuit conductors and equipment. In practice, fuses and circuit breakers will fulfil both of these needs.

Overloads

Overloads are overcurrents, occurring in healthy circuits. They may be caused, for example, by faulty appliances or by connecting too many appliances to a circuit.

Short circuits

A short circuit is the current that will flow when a 'bridge' occurs between live conductors (line-to-neutral for single-phase and line-to-line for three-phase). Prospective short-circuit current is the same, but the term is usually used to signify the value of short circuit at fuse or circuit-breaker positions.

Prospective short-circuit current is of great importance. However, before discussing it or any other overcurrent further, it might be wise to address the subject of fuses, circuit breakers and their characteristics.

Fuses and circuit breakers

As we know, a fuse is the weak link in a circuit, which will break when too much current flows, thus protecting the circuit conductors from damage.

It must be remembered that the priority of a fuse or circuit breaker is to protect the circuit conductors, not the appliance or the user. Calculation of cable size, therefore, automatically involves the correct selection of protective devices.

There are many different types and sizes of fuse, all designed to perform a certain function. The IEE Regulations refer to only four of these: BS 3036, BS 88, BS 1361 and BS 1362 fuses. It is perhaps sensible to include, at this point, circuit breakers to BS 3871 and BS EN 60898 and RCBOs to BS EN 61009.

Fuses

A fuse is simply a device which carries a metal element, usually tinned copper, which will melt and break the circuit when excessive current flows. The three types of fuse are:

1. the rewirable or semi-enclosed fuse
2. the cartridge fuse and fuse link
3. the high-breaking-capacity (HBC) fuse.

The rewirable fuse (BS 3036)

A rewirable fuse consists of a fuse, holder, a fuse element and a fuse carrier (the holder and carrier being made of porcelain or Bakelite). The circuits for which this type of fuse is designed have a colour code, which is marked on the fuse holder and is as follows:

45 A – green
30 A – red
20 A – yellow
15 A – blue
5 A – white

Although this type of fuse is very common in older domestic installations, as it is cheap and easy to repair, it has serious disadvantages:

- The fact that it is repairable enables the wrong size of fuse wire (element) to be used.
- The elements become weak after long usage and may break under normal conditions.
- Normal starting-current surges (e.g. when motors are switched on) are 'seen' by the fuse as an overload and may, therefore, break the circuit.
- The fuse holder and carrier can become damaged as a result of arcing in the event of a heavy overload or short circuit.

Cartridge fuse (BS 1361 and BS 1362)

A cartridge fuse consists of a porcelain tube with metal and caps to which the element is attached. The tube is filled with silica.

These fuses are found generally in modem plug tops used with 13 A socket outlets (BS 1362), in some distribution boards and at mains intake positions (BS 1361). They have the advantage over the rewirable fuse of not deteriorating, of accuracy in breaking at rated values and of not arcing when interrupting faults. They are, however, expensive to replace.

High-breaking-capacity fuses

The HBC fuse is a sophisticated variation of the cartridge fuse and is normally found protecting motor circuits and industrial installations. It consists of a porcelain body filled with silica with a silver element and lug type end caps. Another feature is the indicating element, which shows when the fuse has blown.

It is very fast-acting and can discriminate between a starting surge and an overload.

Circuit breakers

These protective devices have two elements, one thermal and one electro-magnetic. The first, a bi-metal strip, operates for overloads and the second, a sensitive soleniod, detects short circuits.

These devices have the advantage over the fuse in that they may be reset after they have operated (provided the fault current has caused no damage).

Class of protection

It will be evident that each of the protective devices just discussed provides a different level of protection (i.e. rewirable fuses are slower to operate and less accurate than circuit breakers). In order to classify these devices, it is important to have some means of knowing their circuit breaking and fusing performance. This is achieved for fuses by the use of a fusing factor:

$$\text{Fusing factor} = \frac{\text{Fusing current}}{\text{Current rating}}$$

where the fusing current is the minimum current causing the fuse to operate and the current rating is the maximum current which the fuse is designed to carry without operating. For example, a 5 A fuse which operates only when 9 A flows will have a fusing factor of 9/5 = 1.8.

Rewirable fuses have a fusing factor of about 1.8.
Cartridge fuses have a fusing factor of between 1.25 and 1.75.
HBC fuses have a fusing factor of up to 1.25 (maximum).
Circuit breakers are designed to operate at no more than 1.45 times their rating.

Breaking capacity of fuses and circuit breakers

When a short circuit occurs, the current may, for a fraction of a second, reach hundreds or even thousands of amperes. The protective device must be able to break or, in the case of a circuit breaker, make such a current without damage to its surroundings by arcing, overheating or the scattering of hot particles.

Table 5.2 indicates the breaking capacity of fuses typically found in domestic installations.

Table 5.2   Breaking Capacities of Fuses Used in Domestic Installations.

British Standard Number	Current Rating	Breaking Capacity
BS 3036 semi-enclosed re-wireable	5–60 A	up to 4000 A
BS 1362 plug-top cartridge fuse	2–13 A	6000 A
BS 1361 service fuse and consumer unit fuse	5–100 A	16 500 A

The breaking capacity of a circuit breaker to BS 3871 type 1, 2 or 3 of any rating is indicated by an 'M' number (i.e. M3-3 kA, M6-6 kA and M9-9 kA).

For circuit breakers to BS EN 60898 types B, C and D, of any rating, there are two values quoted. The first, **Icn**, must always exceed the value of the prospective short-circuit current at the point at which it is installed. The second, **Ics**, is the maximum fault current the device can withstand without damage or loss of performance. Typical values are

Icn – 1.5 kA, 3 kA, 6 kA, 10 kA, 15 kA, 25 kA
Ics – 1.5 kA, 3 kA, 6 kA, 7.5 kA, 7.5 kA, 10 kA

Circuit breakers to BS 3871 are no longer manufactured, but will, of course, remain in use until they need replacing. Of the BS EN 60898 range, it is the type B that will be most common in domestic installations.

Remember, both fuses and circuit breakers must be selected not only for their ability to carry the design current, but also to break and, in the case of circuit breakers, make the prospective short-circuit current at the point at which they are installed.

Discrimination

In any typical domestic installation there is a main DNO fuse and final circuit protection. In some instances, there is a local plug-top or connection unit fuse for socket outlets and water heating, etc. The ratings of these devices should be such that the higher-rated device should not operate in the event of a fault in part of the installation protected by a lower-rated device. For example, a fault on a washing machine should only operate the 13 A BS 1362 plug-top fuse, not the circuit protection or the main intake fuse.

Position of protective devices

When there is a reduction in the current-carrying capacity of a conductor, a protective device is required. There are, however, some exceptions to this requirement. These are listed clearly in the IEE Regulations. As an example, protection is not needed at a ceiling rose, where the cable size changes from the $1.0\,\text{mm}^2$ to, say, the $0.5\,\text{mm}^2$ for the lampholder flex. This is permitted as it is not expected that lamps will cause overloads.

Residual current devices

No commentary on protective devices would be complete without reference to RCDs. The uses and applications of these devices have already been indicated in Table 3.3, Chapter 2.

Principle of operation of an RCD

Figure 5.1 illustrates the construction of an RCD. In a healthy circuit, the same current passes through the line coil and the load, and then back through the neutral coil. Hence, the magnetic effects of line and neutral currents cancel out. In a faulty circuit,

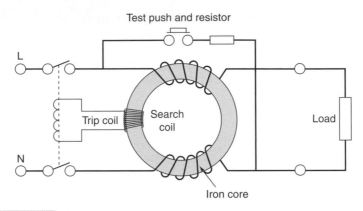

Test push and resistor

L

Trip coil   Search coil   Load

N

Iron core

FIGURE 5.1

either line-to-earth or neutral-to-earth, these currents are no longer equal. Therefore, the out-of-balance current produces some residual magnetism in the core. As this magnetism is alternating, it links with the turns of the search coil, inducing an electro-motive force (EMF) in it. This EMF in turn drives a current through the trip coil, causing operation of the tripping mechanism.

Nuisance tripping

Certain appliances (e.g. cookers, water heaters, etc.) tend to have, by the nature of their construction and use, some leakage currents to earth and freezers cause some switching surges. These are quite normal, but could cause the operation of an RCD protecting an entire installation.

This can be overcome by using split-load consumer units, where socket-outlet circuits are protected by a 30 mA RCD, leaving all other circuits controlled by a normal mains switch, or, as we have seen in Chapter 2 for TT systems, the use of a 100 mA RCD for protecting circuits other than socket outlets.

Probably the best method is to use RCBOs. These are combined RCD/CBs and can be selected in rating for each individual circuit.

CHAPTER 6

Circuit Design

Before we embark on the process of calculating cable sizes, it is perhaps best to consider the common types of accessory and appliance that are used in domestic installations. All current-consuming electrical apparatus is given a rating which indicates the amount of power it will consume.

Some typical loadings of domestic appliances are given in Table 6.1.

Table 6.1   Typical Loadings of Domestic Apparatus.

Appliance	Loading (W)
Lamps (filament)	25, 40, 60, 100, 150
Lamps (fluorescent)	20, 30, 40, 50, 65, 75, 85
Fires and heaters	500–3000
Water heaters, immersion heaters	750–3000
Dishwashers	3000
Washing machines	3000
Spin driers	300–500
Tumble driers	2000–3000
Refrigerators	300–400
Cookers	6000–8000
Portable appliances (e.g. irons, kettles and vacuum cleaners)	10–3000

Conductors and the current-carrying components of accessories (e.g. switches, fuses, circuit breakers, socket outlets and plugs) must be large enough to carry the maximum current which the connected apparatus can cause to flow, without overheating or being overstressed. Conductors and accessories are rated in terms of current in amperes. Therefore, before the required size or 'rating' of a conductor or accessory can be determined, the consumption of the connected apparatus in amperes must be calculated. This is known as the **design current** (I_b) of the circuit and can be determined from

$$I_b = \frac{\text{Power (watts)}}{\text{Volts}}$$

As an example, a 1 kW/230 V electric fire takes 4.35 A when connected to a 230 V supply.

DIVERSITY

The size of a cable or accessory is not necessarily determined by the total power rating of all the current-consuming devices connected to it. It depends on what percentage of the connected load is likely to be operating at any one time. This percentage use is called the **diversity factor**.

Table 6.2 gives an indication of the diversity factors that may be applied to parts of an installation, but it must be remembered that the figures given are only a guide. The amount by which the figures given are increased or decreased for any given installation should be decided by the engineer responsible for the design. The values given in the table refer to percentages of connected load. In calculating the maximum current, appliances and socket outlets should be considered in order of their current ratings, the largest first.

It should be noted that the object of applying diversity to domestic final circuits is not to enable a reduction in cable size, but to arrive at a reduced current demand for the whole installation. This will mean that the size of the main tails, consumer unit and any control gear, etc., can be realistically sized

Example: The maximum demand of an 8 kW/230 V cooker would be

$$8000/230 = 35 \text{ A}$$

Table 6.2   Allowances for Diversity.

Circuit	Percentage Diversity
Lighting	66%
Cooking appliances	the first 10A of the cooker load + 30% of the remainder + 5A if the cooker unit has a socket outlet
Instantaneous water heaters (showers, etc.)	100% of full load of the first and second largest appliances + 25% of full load of remaining appliances
Water heaters thermostatically controlled (immersion heaters)	no diversity allowed
Ring and radial circuits to BS 1363	100% of full load of largest circuit + 40% of full load of all other circuits

The assumed demand after applying diversity would be

$$10 + 30\% \text{ of } (35 - 10) = 10 + 30 \times 25/100 = 17.5 \text{ A}$$

This is less than the current rating of a $1.5\,\text{mm}^2$ cable, but it would not be wise to supply the cooker with that size. The cable size should be based on the cooker's maximum demand.

Where fluorescent or other discharge lighting is involved, a factor of 1.8 is used to take into consideration the associated control gear.

For example, an $80\,\text{W}/230\,\text{V}$ fluorescent fitting will have a current rating of

$$80 \times 1.8/230 = 0.63 \text{ A}$$

BASIC CIRCUIT DESIGN

How do we begin to design? Clearly, plunging into calculations of cable size is of little value unless the type of cable and its method of installation is known. This, in turn, will depend on the installation's environment. At the same time, we would need to know whether the supply was single- or three-phase, what the earthing arrangements were and so on. Here then is our starting point and it is referred to in Part 3 of the IEE Regulations as 'Assessment of General Characteristics'.

Having ascertained all the necessary details, we can decide on an installation method, the type of cable and how we will protect against electric shock and overcurrents. We would now be ready to begin the calculation part of the design procedure. Basically, there are eight stages in such a procedure. These are the same whatever the type of installation, be it a cooker circuit or a steel wire armoured (SWA) cable feeding a remote garage.

Here then are the eight basic steps in a simplified form.

1. Determine the design current I_b.
2. Select the rating of the protection I_n.
3. Select the relevant rating factors (CFs).
4. Divide I_n by the relevant CFs to give tabulated cable current carrying capacity I_t.
5. Choose a cable size to suit I_t.
6. Check the voltage drop.
7. Check for shock risk constraints.
8. Check for thermal constraints.

Let us now examine each stage in detail.

Design current

In many instances, the design current I_b is quoted by the manufacturer, but there are times when it has to be calculated. In this case, the following two formulae are involved (one for single-phase and one for three-phase).

Single-phase $I_b = P/V$ (V is usually 230 V)

Three-phase $I_b = \dfrac{P}{\sqrt{3} \times V_L}$ (V_L is usually 400 V)

Nominal setting of protection I_n

Having determined I_b, we must now select the nominal setting of the protection I_n such that $I_n \geq I_b$. This value may be taken from IEE Regulations or from the manufacturer's charts. The choice of fuse or circuit breaker type is also important and may have to be changed if cable sizes or loop impedances are too high.

Rating factors

When a cable carries its full-load current, it can become warm. This is no problem unless its temperature rises further due to other influences, in which case the insulation could be damaged by overheating. These other influences are high ambient temperature, cables grouped together closely, uncleared overcurrents and contact with thermal insulation. For each of these conditions, there is a rating factor (CF) which will respectively be called C_a, C_g, C_c and C_i, the application of which will have the effect of correcting cable current-carrying capacity or cable size.

Ambient temperature C_a

The cable ratings quoted in the IEE Regulations are based on an ambient temperature of 30°C, and so it is only above this temperature that an adverse correction is needed. The regulations give factors for all types of protection.

Grouping C_g

When cables are grouped together, they impart heat to each other. Therefore, the more cables there are the more heat they will generate, thus increasing the temperature of each cable. The IEE

Regulations give factors for such groups of cables or circuits. It should be noted that the figures given are for cables equally loaded and so correction may not necessarily be needed for cables grouped at the outlet of a domestic consumer unit, for example where there is a mixture of different loads.

Protection by BS 3036 fuse and circuit conditions C_c

As we have already discussed in Chapter 5, BS 3036 fuses have a high fusing factor and, as a result, a factor of 0.725 must always be applied when BS 3036 fuses are used together with any factor for the installation conditions such as routing a cable underground.

Thermal insulation C_i

With the modern trend towards energy saving and the installation of thermal insulation, there may be a need to derate cables to account for heat retention. The values of cable current-carrying capacity given in the IEE Regulations have been adjusted for situations when thermal insulation touches one side of a cable.

However, if a cable is totally surrounded by thermal insulation for more than 0.5 m, a factor of 0.5 must be applied to the tabulated clipped direct ratings. For less than 0.5 m, derating factors shown in Table 6.3 should be applied.

Table 6.3   Thermal Insulation Factors.

Length of Cable (m) in Thermal Insulation	Derating Factor
50	0.88
100	0.78
200	0.63
400	0.51

Application of rating factors

Some or all of the onerous conditions just outlined may affect a cable along its whole length or parts of it at the same time. So, consider the following:

1. If a cable ran for the whole of its length, grouped with others of the same size in a high ambient temperature, and was totally surrounded with thermal insulation, it would seem logical to apply all the CFs, as they all affect the whole cable run. Certainly, the factors for the BS 3036 fuse, grouping and thermal insulation should be used. However, it is doubtful if the ambient temperature will have any effect on the cable, as the thermal insulation, if it is efficient, will prevent heat reaching the cable. Hence, apply C_a, C_g and C_c.

2. If, however, the cable first runs grouped, then leaves the group and runs in high ambient temperature and finally is enclosed in thermal insulation, there will be three different conditions, each affecting the cable in different areas. The BS 3036 fuse affects the whole cable run and, therefore, C_f must be used, but there is no need to apply all of the remaining factors as the worst one will automatically compensate for the others.

Having chosen the relevant rating factors, we now apply them as divisors to the rating of the protective device I_n in order to calculate the tabulated current-carrying capacity I_t of the cable to be used.

Tabulated current-carrying capacity I_t

$$I_t \geq \frac{I_n}{\text{Relevant CFs}}$$

I_n may be replaced by the design current I_b, if the circuit is not likely to be overloaded.

Cable selection

Having established the tabulated current-carrying capacity of the cable required, the actual size is found from a relevant table in the IEE Regulations.

Voltage drop

The resistance of a conductor increases as the length increases and/or the cross-sectional area decreases. Associated with an increased resistance is a drop in voltage, which means that a load at the end of a long, thin cable will not have the full supply voltage available.

The IEE Regulations require that the voltage drop V should not be so excessive that equipment does not function safely. They further indicate that a drop of no more than 3% of the nominal voltage for a lighting circuit and 5% for a power circuit will satisfy. This means that

- for single-phase 230 V, the voltage drop for lighting should not exceed 3% of 230 V = 6.9 V and for power should not exceed 5% of 230 V = 11.5 V
- for three-phase 400 V, the voltage drop should not exceed 12 and 20 V.

For example, the voltage drop on a power circuit supplied from a 230 V source by a 16.0 mm^2 two-core copper cable 23 m long, clipped direct and carrying a design current of 33 A will be:

Cable volt drop $V_c = \dfrac{(mV)}{1000} \times lb \times length(l)/(mv$ from IEE Regs.)

$$= \frac{2.8 \times 33 \times 23}{1000} = 2.125\,V$$

As we have just seen, the maximum volt drop for a 230 V power circuit is 11.5 V, so we can determine the maximum length of the cable by transposing this formula

$$Max\ length = \frac{V_c \times 1000}{mV \times I_b}$$

$$= \frac{11.5 \times 1000}{2.8 \times 33} = 124\,m$$

There are other constraints, however, that may not permit such a length.

Shock risk

This topic has already been discussed in full in Chapter 3. To recap, however, the actual loop impedance Z_s should not exceed those values given in the IEE Regulations.

This ensures that circuits feeding final circuits up to 32 A will be disconnected, in the event of an earth fault, in less than 0.4 s, and that distribution circuits will be disconnected in less than 5 s.

Remember,

$$Z_s = Z_e + (R_1 + R_2)$$

Thermal constraints

The IEE Regulations require that we either select or check the size of a circuit protective conductor (cpc) against tabulated values, or calculate its size using an equation.

Selection of cpc using tabulated values

These simply tell us that

- for line conductors up to and including $16.0\,\mathrm{mm}^2$, the cpc should be at least the same size
- for line conductors between 16.0 and $35.0\,\mathrm{mm}^2$, the cpc should be at least $16\,\mathrm{mm}^2$
- for line conductors over $35.0\,\mathrm{mm}^2$, the cpc should be at least half this size.

This is all very well, but for large sizes of line conductor the cpc is also large and hence costly to supply and install. Also, composite cables, such as the typical twin with cpc 6242 Y type, have cpc's smaller than the line conductor and consequently do not comply with the statements just made.

Calculation of cpc using the adiabatic equation

The adiabatic equation is

$$S = \frac{\sqrt{I^2 \times t}}{k}$$

where:
S = minimum CSA of cpc
I = fault current
t = disconnection time of protection
k = a factor (from the IEE Regulations) dependent on the conductor material and insulation.

This enables us to check on a selected size of cable or on an actual size in a multicore cable. In order to apply the equation, we need first to calculate the earth fault current from

$$I = \frac{U_0}{Z_s}$$

where U_0 is the nominal voltage to earth (usually 230 V) and Z_s is the actual earth fault loop impedance.

Next, we select a k factor from the IEE Regulations and then determine the disconnection time t from the relevant curve (once again in the Regulations).

When we apply these figures to the equation, the value we obtain for 'S' must be less than the size we have chosen or is incorporated in the cable.

An example of circuit design

A consumer lives in a bungalow with a detached garage and workshop, as shown in Figure 6.1. The building method is traditional brick and timber.

The mains intake position is at high level and comprises an 80 A BS 1361 230 V main fuse, an 80 A rated meter and a six-way 80 A consumer unit housing BS EN 60898 Type B circuit breakers as follows:

Ring circuit	32 A
Lighting circuit	6 A
Immersion heater circuit	16 A
Cooker circuit	32 A
Shower circuit	32 A
Spare way	–

The cooker is rated at 30 A, with no socket in the cooker unit. The main tails are 16.0 mm² double-insulated PVC, with a 6.0 mm² earthing conductor. There is no main protective equipotential bonding. The earthing system is TN-S, with an external loop impedance Z_e of 0.3 Ω. The prospective short-circuit current (PSCC) at the origin has been measured as 800 A. The roof space is insulated to full depth of the ceiling joists and the temperature in the roof space is not expected to exceed 35°C.

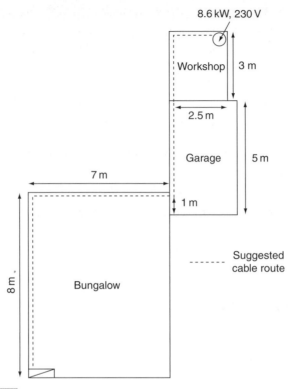

8.6 kW, 230 V

Workshop   3 m

2.5 m

Garage   5 m

7 m

1 m

‑‑‑‑‑‑‑   Suggested
             cable route

8 m

Bungalow

FIGURE 6.1

The consumer wishes to convert the workshop into a pottery room and install an 8.6 kW/230 V electric kiln. The design procedure is as follows.

Assessment of general characteristics

The present maximum demand, applying diversity, is

Ring	32 A
Lighting (66% of 6 A)	3.96 A
Immersion heater	16 A
Cooker (10 A + 30% of 20 A)	16 A
Shower	32 A
Total	**100 A**

Reference to the current rating tables in the IEE Regulations will show that the existing main tails are too small and should be uprated. So, the addition of another 8.6 kW of load is not possible with the present arrangement.

The current taken by the kiln is 8600/230 = 37.4 A. Therefore, the new maximum demand is 100 + 37.4 = 137.4 A.

Supply details are single-phase 230 V, 50 Hz. Earthing: TN-S PSCC at origin (measured): 800 A. Decisions must now be made as to the type of cable, the installation method and the type of protective device. As the existing arrangement is not satisfactory, the supply authority must be informed of the new maximum demand, as a larger main fuse and service cable may be required.

Sizing the main tails

1. The new load on the existing consumer unit will be 137.4 A. From the IEE Regulations, the cable size is 25.0 mm².
2. The earthing conductor size, from the IEE Regulations, will be 16.0 mm². The main protective bonding conductor size, from the IEE Regulations, will be 10.0 mm².

For a domestic installation such as this, a PVC flat twin-cable, clipped direct (avoiding any thermal insulation) through the loft space and the garage, etc., would be most appropriate.

Sizing the kiln circuit cable

Design current I_b

$$I_b = \frac{P}{V} = \frac{8600}{230} = 37.4 \text{ A}$$

Rating and type of protection I_n

As we have seen, the requirement for the rating I_n is that $I_n \geq I_b$. Therefore, using the tables in the IEE Regulations, I_n will be 40 A.

Rating factors

C_a – 0.94
C_g – not applicable
C_c – 0.725 **only** if the fuse is BS 3036 (not applicable here)
C_i – 0.5 if the cable is totally surrounded in thermal insulation (not applicable here).

Tabulated current-carrying capacity of cable

$$I_t = \frac{I_n}{\text{CF}} = \frac{40}{0.94} = 42.5 \text{ A}$$

Cable size based on tabulated current-carrying capacity.

Table 4D5A IEE Regulations gives a size of 6.0 mm² for this value of I_t (method C).

Check on voltage drop

The actual voltage drop is given by

$$\frac{\text{mV} \times I_b \times 1}{1000} = \frac{7.3 \times 37.4 \times 24.5}{1000} = 6.7 \text{ V}$$

(well below the maximum of 11.5 V)

This voltage drop, whilst not causing the kiln to work unsafely, may mean inefficiency, and it is perhaps better to use a 10.0 mm² cable.

For a 10.0 mm² cable, the voltage drop is checked as

$$\frac{4.4 \times 37.4 \times 24.5}{1000} = 4.04 \text{ V}$$

Shock risk

The cpc associated with a $10.0\,\text{mm}^2$ twin 6242 Y cable is $4.0\,\text{mm}^2$. Hence, the total loop impedance will be

$$Z_\text{s} = Z_\text{e} + \frac{(R_1 + R_2) \times L \times 1.2}{1000}$$

$$= 0.3 + \frac{6.44 \times 24.5 \times 1.2}{1000} = 0.489\,\Omega$$

Note

6.44 is the tabulated $(R_1 + R_2)$ value and the multiplier 1.2 takes account of the conductor resistance at its operating temperature.

It is likely that the chosen circuit breaker will be a type B.

Thermal constraints

We still need to check that the $4.0\,\text{mm}^2$ cpc is large enough to withstand damage under earth fault conditions. So, the fault current would be

$$I = \frac{U_0}{Z_\text{s}} = \frac{230}{0.489} = 470\,\text{A}$$

The disconnection time t for this current for this type of protection (from the relevant curve in the IEE Regulations) is as follows.

40 A circuit breaker type B $= 0.1\,\text{s}$ (the actual time is less than this but $0.1\,\text{s}$ is the instantaneous time)

From the regulations, the factor for $k = 115$. We can now apply the adiabatic equation

$$S = \frac{\sqrt{I^2 \times t}}{k} = \frac{\sqrt{470^2 \times 0.1}}{115} = 1.29\,\text{mm}^2$$

Hence, our $4.0\,\text{mm}^2$ cpc is of adequate size.

SUMMARY

- The kiln circuit would be protected by a 40 A BS EN 60898 type B circuit breaker and supplied from a spare way in the consumer unit using $10.0\,\text{mm}^2$ twin with earth PVC cable.
- The main fuse would need to be uprated to 100 A.
- The main tails would be changed to $25.0\,\text{mm}^2$.
- The earthing conductor would be changed to $16.0\,\text{mm}^2$.
- Main protective bonding conductors would need to be installed.

CHAPTER 7

Inspection and Testing

Apart from the knowledge required to carry out the verification process competently, the person conducting the inspection and test must be in possession of appropriate test instruments.

INSTRUMENTS

In order to fulfil the basic requirements for testing to the IEE Regulations (BS 7671), the following instruments are needed:

- a continuity tester (low ohms)
- an insulation resistance tester
- a loop impedance tester
- a residual current device (RCD) tester
- a prospective fault current (PFC) tester
- an approved test lamp or voltage indicator
- a proving unit.

Many instrument manufacturers have developed dual or multi-function instruments, and so it is quite common to have continuity and insulation resistance in one unit, loop impedance and PFC in one unit, and loop impedance, PFC and RCD tests in one unit, etc. However, regardless of the various combinations, let us take a closer look at the individual test instrument requirements.

A continuity tester (low resistance ohmmeter)

Bells, buzzers and simple multimeters will all indicate whether or not a circuit is continuous, but will not show the difference

105

between the resistance of, say, a 10 m length of 10.0 mm² conductor and a 10 m length of 1.0 mm² conductor.

A continuity tester should have a no-load source voltage of between 4 and 24 V, and be capable of delivering an AC or DC short-circuit voltage of not less than 200 mA. It should have a resolution (i.e. a detectable difference in resistance) of at least 0.05 mΩ.

An insulation resistance tester

An insulation resistance tester must be capable of delivering 1 mA when the required test voltage is applied across the minimum acceptable value of insulation resistance.

Consequently, an instrument selected for use on a low-voltage system should be capable of delivering 1 mA at 500 V across a resistance of 1 mΩ.

A loop impedance tester

This instrument functions by creating, in effect, an earth fault for a brief moment, and is connected to the circuit via a plug or by 'flying leads' connected separately to line, neutral and earth.

The instrument should only allow an earth fault to exist for a maximum of 40 ms and a resolution of 0.01 Ω is adequate for circuits up to 50 A. Above this circuit rating, the ohmic values become too small to give such an accuracy using a standard instrument and more specialized equipment may be required.

An RCD tester

Usually connected by the use of a plug, although flying leads are needed for non-socket-outlet circuits, this instrument allows a range of out-of-balance currents to flow through the RCD to cause its operation within specified time limits.

The test instrument should not be operated for longer than 2 s and it should have a 10% accuracy across the full range of test currents.

A PFC tester

Normally one half of a dual, loop impedance/PSCC tester, this instrument measures the prospective line-neutral or line-earth fault current at the point of measurement using the flying lead.

An approved test lamp or voltage indicator

A flexible cord with a lamp attached is not an approved device, nor for that matter is the ubiquitous 'testascope' or 'neon screw-driver', which encourages the passage of current, at low voltage, through the body!

A typical approved test lamp is as shown in Figure 7.1.

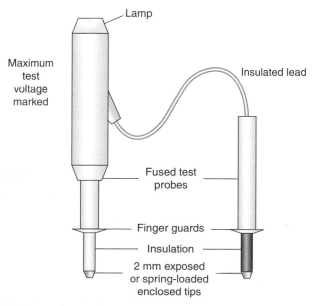

FIGURE 7.1 Approved test lamp.

The Health and Safety Executive, Guidance Note 38, recommends that the leads and probes associated with test lamps, voltage indicators, voltmeters, etc., have the following characteristics:

1. The leads should be adequately insulated and, ideally, fused.
2. The leads, where separate, should be easily distinguished from each other by colour.
3. The leads should be flexible and sufficiently long for their purpose.
4. The probes should incorporate finger barriers to prevent accidental contact with live parts.
5. The probes should be insulated and have a maximum of 2 mm of exposed metal, but preferably have spring-loaded enclosed tips.

A proving unit

This is an optional item of test equipment, in that test lamps should be proved on a known supply which could, of course, be an adjacent socket or lighting point. However, to prove a test lamp on such a known supply may involve entry into enclosures with the associated hazards that could bring. A proving unit is a compact device, not much larger than a pack of cards, which is capable of electronically developing 230 V DC across which the test lamp may be proved. The exception to this are test lamps incorporating 230 V lamps which will not activate from the small power source of the proving unit.

Test lamps must be proved against a voltage similar to that to be tested. So, proving test lamps that incorporate an internal check (i.e. shorting out the probes to make a buzzer sound) is not acceptable if the voltage to be tested is higher than that delivered by the test lamp.

Care of test instruments

The 1989 Electricity at Work Regulations require that all electrical systems (including test instruments) be maintained to prevent danger. This does not restrict such maintenance to just a yearly calibration, but requires equipment to be kept in good condition and regularly checked for accuracy against known values (e.g. a checkbox). In this way they will be safe to use at all times.

Whilst test instruments and associated leads, probes and clips used in the electrical contracting industry are robust in design and manufacture, they still need to be treated with care and protected from mechanical damage.

Test gear should be kept in a separate box or case away from tools and sharp objects and the general condition of a tester and leads should always be checked before they are used.

Initial inspection

Before any testing is carried out, a detailed physical inspection must be made to ensure that all equipment is

- to a relevant British or Harmonized European Standard
- erected/installed in compliance with the IEE Regulations
- not damaged in such a way that it could cause danger.

In order to comply with these requirements, the Regulations give a checklist of items that, where relevant, should be inspected.

However, before such an inspection, and test for that matter, is carried out, certain information **must** be available to the verifier.

This information is the result of the 'Assessment of General Characteristics' required by IEE Regulations (Part 3, sections 311, 312 and 313), together with drawings, charts and similar information relating to the installation. For domestic installations, these diagrams and charts are usually in the form of a simple schedule.

Interestingly, one of the items on the checklist **is** the presence of diagrams, instructions and similar information. If these are missing, there is a deviation from the Regulations.

Another item on the list is the verification of conductors for current-carrying capacity and voltage drop in accordance with the design. How could this be verified without all the information? A 32 A type B circuit breaker protecting a length of $4.0\,\mathrm{mm}^2$ conductor may look reasonable, but is it correct and, unless you are sure, are you prepared to sign to say that it is?

Let us look then at the general content of the checklist:

1. *Connection of conductors:* Are terminations electrically and mechanically sound? Is insulation and sheathing removed only to a minimum to allow satisfactory termination?

2. *Identification of conductors:* Are conductors correctly identified in accordance with the regulations?

3. *Routing of cables:* Are cables installed such that account is taken of external influences, such as mechanical damage, corrosion and heat?

4. *Conductor selection:* Are conductors selected for current-carrying capacity and voltage drop in accordance with the design?

5. *Connection of single pole devices:* Are single pole protective and switching devices connected in the line conductor only?

6. *Accessories and equipment:* Are all accessories and items of equipment correctly connected?

7. *Thermal effects:* Are fire barriers present where required and protection against thermal effects provided?

8. *Protection against shock:* What methods have been used to provide protection against electric shock?

9. *Mutual detrimental influence:* Are wiring systems installed so that they can have no harmful effect on non-electrical systems or so that systems of different currents or voltages are segregated where necessary?

10. *Isolation and switching:* Are the appropriate devices for isolation and switching correctly located and installed?

11. *Undervoltage:* Where undervoltage may give rise for concern, are there protective devices present?

12. *Protective devices:* Are protective and monitoring devices correctly chosen and set to ensure automatic disconnection and/or overcurrent?

13. *Labelling:* Are all protective devices, switches (where necessary) and terminals correctly labelled?

14. *External influences:* Have all items of equipment and protective measures been selected in accordance with the appropriate external influences?

15. *Access:* Are all means of access to switchgear and equipment adequate?

16. *Notices and signs:* Are danger notices and warning signs present?

17. *Diagrams:* Are diagrams, instructions and similar information relating to the installation available?

18. *Erection methods:* Have all wiring systems, accessories and equipment been selected and installed in accordance with the requirements of the Regulations? Are fixings for equipment adequate for the environment?

Once all the relevant items have been inspected, providing there are no defects that may lead to a dangerous situation when testing, the actual testing procedure can start.

TESTING

Continuity of protective conductors

All protective conductors, including main and supplementary bonding conductors must be tested for continuity using a low-reading ohmmeter.

For main equipotential bonding, there is no single fixed value of resistance above which the conductor would be deemed unsuitable. Each measured value, if indeed it is measurable for very short lengths, should be compared with the relevant value for a particular conductor length and size. Such values are shown in Table 7.1.

Table 7.1   Resistance (Ω) of Copper Conductors at 20°C.

CSA (mm)	Length (m)									
	5	10	15	20	25	30	35	40	45	50
1	0.9	0.18	0.27	0.36	0.45	0.54	0.63	0.72	0.82	0.9
1.5	0.06	0.12	0.18	0.24	0.3	0.36	0.43	0.48	0.55	0.6
2.5	0.04	0.07	0.11	0.15	0.19	0.22	0.26	0.03	0.33	0.37
4	0.023	0.05	0.07	0.09	0.12	0.14	0.16	0.18	0.21	0.23
6	0.02	0.03	0.05	0.06	0.08	0.09	0.11	0.13	0.14	0.16
10	0.01	0.02	0.03	0.04	0.05	0.06	0.063	0.07	0.08	0.09
16	0.006	0.01	0.02	0.023	0.03	0.034	0.04	0.05	0.05	0.06
25	0.004	0.007	0.01	0.015	0.02	0.022	0.026	0.03	0.033	0.04
35	0.003	0.005	0.008	0.01	0.013	0.016	0.019	0.02	0.024	0.03

Where a supplementary protective bonding conductor has been installed between **simultaneously accessible** exposed and extraneous conductive parts the resistance of the conductor R must be equal to or less than $50/I_a$.

So, $R \leq 50/I_a$, where 50 is the voltage, above which exposed metalwork should not rise and I_a is the minimum current, causing operation of the circuit protective device within 5 s.

For example, suppose a 45 A BS 3036 fuse protects a cooker circuit, the disconnection time for the circuit cannot be met, and so a supplementary bonding conductor has been installed between the cooker case and the adjacent metal sink. The resistance R of that conductor should not be greater than $50/I_a$, which in this case is 145 A (IEE Regulations). So

$$50/145 = 0.34\ \Omega$$

How then do we conduct a test to establish continuity of main or supplementary bonding conductors? Quite simple really, just connect the leads from the continuity tester to the ends of the bonding conductor (Figure 7.2). One end should be disconnected from its bonding clamp, otherwise any measurement may include the resistance of parallel paths of other earthed metalwork. Remember to zero the instrument first or, if this facility is not available, record the resistance of the test leads so that this value can be subtracted from the test reading.

Important Note

If the installation is in operation, **never** disconnect main bonding conductors unless the supply can be isolated. Without isolation, persons and livestock are at risk of electric shock.

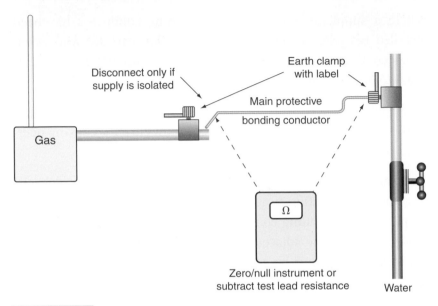

Disconnect only if
supply is isolated

Earth clamp
with label

Main protective
bonding conductor

Gas

Ω

Zero/null instrument or
subtract test lead resistance

Water

FIGURE 7.2 Continuity of main protective bonding conductors.

The continuity of circuit protective conductors (cpc's) may be established in the same way, but a second method is preferable, as the results of this second test indicate the value of $(R_1 + R_2)$ for the circuit in question.

The test is conducted in the following way (Figure 7.3):

1. Temporarily link together the line conductor and cpc of the circuit concerned in the distribution board or consumer unit.
2. Test between line and cpc at each outlet in the circuit. A reading indicates continuity.
3. Record the test result obtained at the furthest point in the circuit. This value is $(R_1 + R_2)$ for the circuit.

There may be some difficulty in determining the $(R_1 + R_2)$ values of circuits in installations that comprise steel conduit and trunking and/or SWA and MIMS cables because of the parallel earth paths

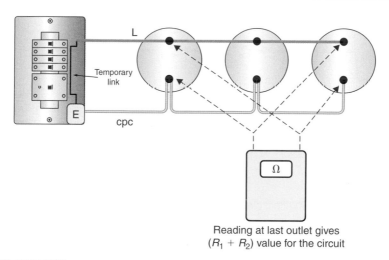

Reading at last outlet gives
$(R_1 + R_2)$ value for the circuit

FIGURE 7.3   CPC continuity.

that are likely to exist. In these cases, continuity tests may have to be carried out at the installation stage before accessories are connected or terminations made off as well as after completion.

Continuity of ring final circuit conductors

The two main reasons for conducting this test are:

1. to establish that interconnections in the ring do not exist
2. to ensure that the cpc is continuous and to indicate the value of $(R_1 + R_2)$ for the ring.

What then are interconnections in a ring circuit and why is it important to locate them? Figure 7.4 shows a ring final circuit with an interconnection.

The most likely cause of this situation is where a DIY enthusiast has added sockets P, Q, R and S to an existing ring A, B, C, D, E and F.

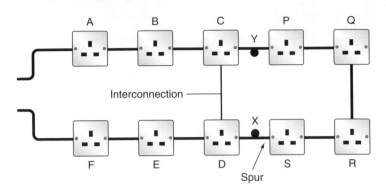

FIGURE 7.4 Ring final circuit with inter-connection.

In itself, there is nothing wrong with this. The problem arises if a break occurs at, say, point Y, or if the terminations fail in socket C or P. Then there would be four sockets all fed from the point X, which would then become a spur.

So, how do we identify such a situation with or without breaks at point Y? A simple resistance test between the ends of the line, neutral or cpc's will only indicate that a circuit exists, whether there are interconnections or not. The following test method is based on the idea that the resistance measured across any diameter of a perfect circle of conductor will always be the same value (Figure 7.5).

The perfect circle of conductor is achieved by cross-connecting the line and neutral legs of the ring (Figure 7.6).

The test procedure is as follows:

1. Identify the opposite legs of the ring. This is quite easy with sheathed cables, but with singles, each conductor will have to be identified, probably by taking resistance measurements between each one and the closest socket outlet. This will give three high readings and three low readings, thus establishing the opposite legs.

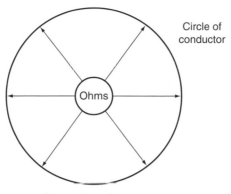

Circle of conductor

Same value whatever diameter is measured

FIGURE 7.5

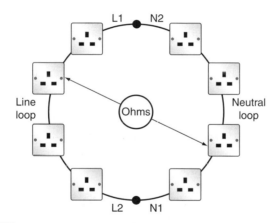

FIGURE 7.6 Circle formed by cross connection.

2. Take a resistance measurement between the ends of each conductor loop. Record this value.
3. Cross-connect the ends of the line and neutral loops (Figure 7.7).
4. Measure between line and neutral at each socket on the ring. For a perfect ring, the readings obtained should be

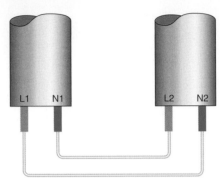

FIGURE 7.7   L and N cross connection.

substantially the same. If an interconnection existed such as shown in Figure 7.4 then sockets A to F would all have similar readings, and those beyond the interconnection would have gradually increasing values to approximately the mid-point of the ring, then decreasing values back towards the interconnection. If a break had occurred at point Y the readings from socket S would increase to a maximum at socket P. One or two high readings are likely to indicate either loose connections or spurs. A null reading, i.e. an open circuit indication, is probably a reverse polarity, either line-cpc or neutral-cpc reversal. These faults would clearly be rectified and the test at the suspect socket(s) repeated.

5. Repeat the above procedure, but in this case cross-connect the line and cpc loops (Figure 7.8).

In this instance, if the cable is of the flat twin type, the readings at each socket will very slightly increase and then decrease around the ring. This difference, due to the line and cpc being different sizes, will not be significant enough to cause any concern. The measured value is very important, it is $R_1 + R_2$ for the ring.

As before, loose connections, spurs and, in this case, L–N cross polarity will be picked up.

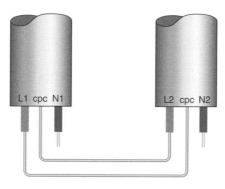

FIGURE 7.8   L and cpc cross connection.

Table 7.2

Initial measurements	L1 to L2	N1 to N2	cpc 1 to cpc 2
	0.52	0.52	0.85
Reading at each socket	0.26	0.26	0.30–0.34
For spurs, each metre in length will add the following resistance to the above values	0.015	0.015	0.02

Table 7.2 gives the typical approximate ohmic values for a healthy 70 m ring final circuit wired in 2.5/1.5 flat twin and cpc cable.

Insulation resistance

This test is probably the most used and yet abused of them all. Affectionately known as 'meggering', an insulation resistance test is carried out to ensure that the insulation of conductors, accessories and equipment is in a healthy condition and will prevent dangerous leakage currents between live conductors and between live conductors and earth. It also indicates whether any short circuits exist.

Insulation resistance, as discussed, is the resistance measured between conductors and is made up of countless millions of resistances in parallel (Figure 7.9).

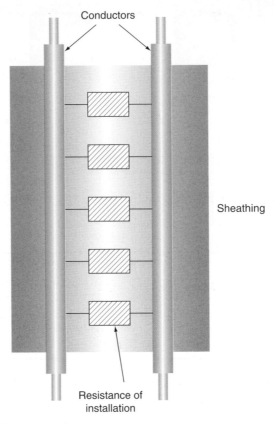

FIGURE 7.9   Cable insulation resistance.

The more resistances there are in parallel, the lower the overall resistance, and, in consequence, the longer a cable the lower the insulation resistance. Add to this the fact that almost all installation circuits are also wired in parallel, and it becomes apparent that tests on large installations may give, if measured as a whole, pessimistically low values, even if there are no faults.

Under these circumstances, it is usual to break down such large installations into smaller sections, floor by floor, distribution circuit by distribution circuit, etc. This also helps, in the case of periodic testing, to minimize disruption. The test procedure then is as follows:

1. Disconnect all items of equipment such as capacitors and indicator lamps, as these are likely to give misleading results. Remove any items of equipment likely to be damaged by the test (e.g. dimmer switches and electronic timers). Remove all lamps and accessories and disconnect fluorescent and discharge fittings. Ensure that the installation is disconnected from the supply, all fuses are in place and circuit breakers and switches are in the on position. In some instances, it may be impracticable to remove lamps, etc., in that case, the local switch controlling such equipment may be left in the off position.

2. Join together all live conductors of the supply and test between this join and earth. Alternatively, test between each live conductor and earth in turn.

3. Test between line and neutral. For three-phase systems, join together all line conductors and test between this join and neutral. Then test between each of the lines. Alternatively, test between each of the live conductors in turn. Installations incorporating two-way lighting systems should be tested twice with the two-way switches in alternate positions.

Table 7.3 gives the test voltages and minimum values of insulation resistance for ELV and LV systems.

Table 7.3

System	Test Voltage	Minimum Insulation Resistance
SELV and PELV	250 V DC	0.5 MΩ
LV up to 500 V	500 V DC	1.0 MΩ
Over 500 V	1000 V DC	1.0 MΩ

If a value of less than $2\,M\Omega$ is recorded, it may indicate a situation where a fault is developing, but as yet still complies with the minimum permissible value. In this case, each circuit should be tested separately to identify any faulty circuits.

Polarity

This simple test, often overlooked, is just as important as all the others and many serious injuries and electrocutions could have been prevented if only polarity checks had been carried out.

The requirements are that

- all fuses and single pole switches are in the line conductor
- the centre contact of an Edison screw type lampholder is connected to the line conductor with the exception of the new E14 and E27 holders which have an insulating material as a screwed part
- all socket outlets and similar accessories are wired correctly.

Although polarity is towards the end of the recommended test sequence, it would seem sensible, on lighting circuits, for example, to conduct this test at the same time as that for continuity of cpc's (Figure 7.10).

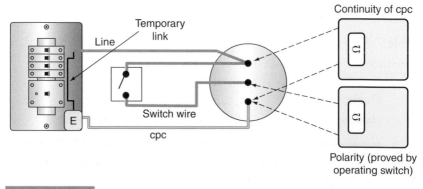

FIGURE 7.10   Lighting polarity.

As discussed, polarity on ring final circuit conductors is achieved simply by conducting the ring circuit test. For radial socket outlet circuits, however, this is a little more difficult. The continuity of the cpc will have already been proved by linking line and cpc and measuring between the same terminals at each socket. Whilst a line-cpc reversal would not have shown, a line-neutral reversal would have, as there would have been no reading at the socket in question. This would have been remedied, and so only line-cpc reversals need to be checked. This can be done by linking together line and neutral at the origin and testing between the same terminals at each socket. A line-cpc reversal will result in no reading at the socket.

Earth fault loop impedance

As we have already seen in Chapter 3, overcurrent protective devices must, under earth fault conditions, disconnect fast enough to reduce the risk of electric shock. This is achieved if the actual value of the earth fault loop impedance does not exceed the tabulated maximum values given in the IEE Regulations.

The purpose of the test, therefore, is to determine the actual value of the loop impedance Z_s, for comparison with those maximum values. It is conducted as follows:

1. Ensure that all main protective bonding is in place.
2. Connect the test instrument either by its BS 1363 plug or the flying leads, to the line, neutral and earth terminals at the remote end of the circuit under test. (If a neutral is not available, e.g. in the case of a three-phase motor, connect the neutral probe to earth.)
3. Press to test and record the value indicated.

It must be understood that this instrument reading is **not valid for direct comparison with the tabulated maximum values**, as account must be taken of the ambient temperature at the time of test and the maximum conductor operating temperature, both of which will have an effect on conductor resistance. Hence, the $(R_1 + R_2)$ value is likely to be greater at the time of the fault than at the time of the test.

So, the measured value of Z_s may be compared with the maximum value by carrying out a calculation involving various correction factors. This method is lengthy and unlikely to be used in general situations. The following rule of thumb method is preferred.

The tabulated maximum value of Z_s is multiplied by 0.8 and the measured value compared with this corrected value

$$\text{Compare } Z_s \text{ measured with } Z_s \text{ maximum} \times 0.8$$

Table 7.4 gives the 0.8 values of tabulated loop impedance for direct comparison with measured values.

In effect, a loop impedance test places a line-earth fault on the installation and, if an RCD is present, it may not be possible to conduct the test, as the device will trip out each time the loop impedance tester button is pressed.

Unless the instrument is of a type that has a built-in guard against such tripping, the value of Z_s will have to be determined from measured values of Z_e and $(R_1 + R_2)$.

Note

Never short out an RCD in order to conduct this test.

Table 7.4 Values of Loop Impedance for Comparison with Test Readings.

Protection	Disconnection Time		5A	6A	10A	15A	16A	20A	25A	30A	32A	40A	45A	50A	60A	63A	80A	100A	125A	160A
BS 3036 fuse	0.4 s	Z_s max	7.6	—	—	—	—	1.41	—	0.87	—	—	—	—	—	—	—	—	—	—
	5 s	Z_s max	14.16	—	—	4.2	—	3.06	—	2.11	—	—	1.27	—	0.89	—	—	—	—	—
BS 88 fuse	0.4 s	Z_s max	—	6.82	4.09	—	2.16	1.42	1.15	—	0.83	—	—	—	—	—	—	—	—	—
	5 s	Z_s max	—	10.8	5.94	—	3.33	2.32	1.84	—	1.47	1.08	—	0.83	—	0.67	0.45	0.33	0.26	0.2
BS 1361 fuse	0.4 s	Z_s max	8.36	—	—	2.62	—	1.36	—	0.92	—	—	—	—	—	—	—	—	—	—
	5 s	Z_s max	13.12	—	—	4	—	2.24	—	1.47	—	—	0.75	—	0.56	—	0.4	0.29	—	—
BS 1362 fuses	0.4 s	Z_s max	(3A) 13.12	—	—	(13A) 1.9	—	—	—	—	—	—	—	—	—	—	—	—	—	—
	5 s	Z_s max	(3A) 18.56	—	—	(13A) 3.06	—	—	—	—	—	—	—	—	—	—	—	—	—	—
BS 3871 MCB Type 1	0.4 & 5 s	Z_s max	9.2	7.6	4.6	3.06	2.87	2.3	1.84	1.53	1.44	1.15	1.02	0.92	—	0.73	—	—	—	—
BS 3871 MCB Type 2	0.4 & 5 s	Z_s max	5.25	4.37	2.62	1.75	1.64	1.31	1.05	0.87	0.82	0.67	0.58	0.52	—	0.42	—	—	—	—
BS 3871 MCB Type 3	0.4 & 5 s	Z_s max	3.68	3	1.84	1.22	1.15	0.92	0.74	0.61	0.57	0.46	0.41	0.37	—	0.29	—	—	—	—
BS EN 60898 CB Type B	0.4 & 5 s	Z_s max	(3A) 12.26	6.13	3.68	—	2.3	1.84	1.47	—	1.15	0.92	—	0.74	—	0.58	0.46	0.37	0.3	—
BS EN 60898 CB Type C	0.4 & 5 s	Z_s max		3.06	1.84	—	1.15	0.92	0.75	—	0.57	0.46	—	0.37	—	0.288	0.23	0.18	0.15	—
BS EN 60898 CB Type D	0.4 & 5 s	Z_s max		1.54	0.92	—	0.57	0.46	0.37	—	0.288	0.23	—	0.18	—	0.14	0.12	0.09	0.07	—

As a loop impedance test creates a high earth fault current, albeit for a short space of time, some lower-rated circuit breakers may operate, resulting in the same situation as with an RCD, and Z_s will have to be calculated. It is not really good practice to temporarily replace the circuit breaker with one of a higher rating.

External loop impedance Z_e

The value of Z_e is measured at the intake position on the supply side and with all main protective bonding disconnected. Unless the installation can be isolated from the supply, this test should not be carried out, as a potential shock risk will exist with the supply on and this bonding disconnected.

Earth electrode resistance

As we know, in many rural areas, the supply system is TT and so reliance is placed on the general mass of earth for a return path under earth fault conditions and connection to earth is made by an electrode, usually of the rod type.

In order to determine the resistance of the earth return path, it is necessary to measure the resistance that the electrode has with earth.

An earth fault loop impedance test is carried out between the incoming line terminal and the electrode (a standard test for Z_e).

The value obtained is added to the cpc resistance of the protected circuits and this value is multiplied by the operating current of the RCD. The resulting value should not exceed 50 V.

Functional testing

RCD operation

Where RCDs are fitted, it is essential that they operate within set parameters. The RCD testers used are designed to do just this. The basic tests required are as follows (Table 7.5 gives further details).

Table 7.5

RCD Type	Half Rated	Full Trip Current
BS 4293 and BS 7288 sockets	No trip	Less than 200 ms
BS 4293 with time delay	No trip	1/2 time delay + 200 ms
BS EN 61009 or BS EN 61009 RCBO	No trip	300 ms
As above, type S with time delay	No trip	130–500 ms

1. Set the test instrument to the rating of the RCD.
2. Set the test instrument to half-rated trip.
3. Operate the instrument and the RCD should not trip.
4. Set the instrument to deliver the full-rated tripping current of the RCD.
5. Operate the instrument and the RCD should trip out in the required time.

When an RCD is used for additional protection against shock, it must be rated at 30 mA or less and operate within 40 ms when subjected to a tripping current of five times its rating $(I_{\Delta n})$.

Test instruments have the facility to provide this value of tripping current. There is no point in conducting this 'fast trip' test if an RCD has a rating in excess of 30 mA.

All RCDs have a built-in test facility in the form of a test button. Operating this test facility creates an artificial out-of-balance condition that causes the device to trip. This only checks the mechanics of the tripping operation, it is not a substitute for the instrument tests.

All other items of equipment such as switchgear, controlgear interlocks, etc., must be checked to ensure that they are correctly mounted and adjusted and that they function correctly.

Periodic inspection

This could be so simple. As it is, periodic inspection and testing tends to be complicated and frustrating. On the domestic scene, few if any house owners actually decide to have a regular inspection. The usual response is 'If it works, it must be OK'. It is usually only when there is a change of ownership that the mortgage companies insist on an electrical survey.

Let us assume that the original installation was erected in accordance with the IEE Wiring Regulations, and that any alterations and/or additions have been faithfully recorded on the original documentation (which is, of course, readily available!!).

A periodic inspection and test under these circumstances should be relatively easy, as little dismantling of the installation will be necessary and the bulk of the work will be inspection.

Inspection should be carried out with the supply disconnected as it may be necessary to gain access to wiring in enclosures, etc. So, with large installations, it will probably need considerable liaison with the client to arrange convenient times for interruption of supplies to various parts of the installation.

This is also the case when testing protective conductors as these **must never** be disconnected unless the supply can be isolated. This is particularly important for main protective bonding conductors which need to be disconnected in order to measure Z_e.

In the main an inspection should reveal

- any aspects of the installation that may impair the safety of persons and livestock against the effects of electric shock and burns
- that there are no installation defects that could give rise to heat and fire and so damage property

- that the installation is not damaged or deteriorated so as to impair safety
- that any defects or non-compliance with the regulations, which may give rise to danger, are identified.

As mentioned earlier, dismantling should be kept to a minimum and, as a result, a certain amount of sampling will take place. This sampling would need to be increased if defects were found.

From the testing point of view, not all of the tests carried out on the initial inspection may need to be applied. This decision depends on the condition of the installation.

The continuity of protective conductors is clearly important, as is insulation resistance and loop impedance, but one wonders if polarity tests are necessary if the installation has remained undisturbed since the last inspection. The same applies to ring circuit continuity as the L–N test is applied to detect interconnections in the ring, which would not happen on their own!

Certification

Having completed all the inspection checks and carried out all the relevant tests, this information needs to be documented. This is done on Electrical Installation Certificates (EICs), inspections schedules, test result schedules, periodic inspection and test reports, minor works certificates and any other documentation you wish to add. Examples of this documentation are shown in the IEE Regulations and the IEE Guidance Note 3 on inspection and testing.

This documentation is vital. It has to be correct and signed by a **competent** person. EICs and periodic test reports **must** be accompanied by a schedule of test results and an inspection schedule for them to be valid. Three signatures are needed on an EIC, one in respect of the design, one in respect of the construction and one in respect of the inspection and test. It could be, of course, that for a very small company, one person signs all three parts. Whatever the case, the original must be given to the person ordering the work and a duplicate retained by the contractor.

One important aspect of the EIC is the recommended interval between inspections. This should be evaluated by the designer and will depend on the type of installation and its usage. In some cases, the time interval is mandatory, especially where environments are subject to use by the public. The IEE Guidance Note 3 gives the recommended maximum frequencies between inspections.

A periodic test report form is very similar in part to an EIC as regards the details of the installation (i.e. maximum demand, type of earthing system and Z_e). The rest of the form deals with the extent and limitations of the inspection and test, the recommendations and a summary of the installation. The record of the extent and limitations of the inspection is very important. It must be agreed with the client or other third party exactly what parts of the installation will be covered by the report and what parts won't.

With regards to the schedule of test results, test values should be recorded **unadjusted**, any compensation for temperature, etc., being made after the testing is completed.

Any alterations or additions to an installation will be subject to the issue of an EIC, except where the addition is, say, a single point added to an existing circuit; then the work is subject to the issue of a minor works certificate.

CHAPTER 8

Security Alarm and Telephone Systems

SECURITY ALARMS

There are two types of intruder alarm system available, the hard-wired system and the wireless variety. Clearly, the latter appears attractive from an installation point of view, as there are no cables to be run. However, nuisance operation can be caused by stray radio frequencies unrelated to the system. The hard-wired type is preferred as it is very reliable.

Simple systems

Most domestic premises require only the most basic of systems, usually comprising an electronically operated control panel, a number of sensors (either passive infrared (PIR) or magnetic) and a sounder (bell or siren). Operation of any sensor is detected by the control panel and the sounder activated. Figure 8.1 shows a block diagram of this set-up.

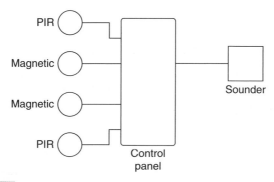

FIGURE 8.1 Alarm system block diagram.

Most systems are of the 'closed-circuit' type, in which the sensors have 'normally closed' (N/C) contacts. Operation of a sensor opens these contacts and the alarm sounder is activated. Cutting through cables has the same effect as operation of a sensor.

Sensors

PIR units react to body heat and movement and require a permanent 9 V DC supply from the control panel as well as battery backup in case of mains failure. They are used to protect areas from intrusion from several directions. Careful consideration must be given to the siting of PIRs in order to gain the best possible protection.

Magnetic sensors require no supply, as they are simply a pair of contacts held closed by the proximity of a magnet. The units housing the contacts are installed in door or window frames and the magnets in the moving parts.

Control panel

There are various models to choose from but, essentially, they all perform the same task. Most panels used in the domestic situation have the facility to protect various zones independent of the others. There are generally four zones catered for.

Panels are supplied from the low-voltage electrical installation usually from a fused connection unit, and they incorporate a battery for continuation of operation in the event of a mains' failure.

Wiring

Wiring can be flush or surface and is usually 7/0.2 mm². PIRs require six cores, magnetic sensors require only two cores. The N/C contacts of sensors are wired in series. Figure 8.2 illustrates how sensors are connected.

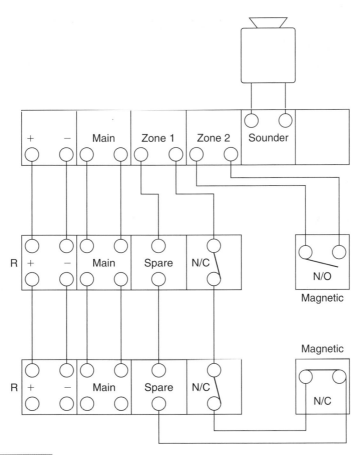

FIGURE 8.2 Sensor connections.

TELEPHONE SYSTEMS

Extensions to the domestic telephone system are extremely easy as each extension socket is wired in parallel with the one previous (Figure 8.3).

The master socket is the first socket in any installation and contains components to

- allow telephones to be removed without causing problems at the exchange

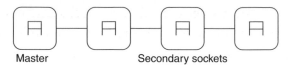

Master   Secondary sockets

FIGURE 8.3 Telephone block diagram.

- stop surges on the line such as lightning strikes
- prevent the telephone making partial ringing noises when dialling.

Connection to the master socket is not permitted, except by use of an adaptor plug and extension cable.

Extension or secondary sockets house only terminals.

Secondary sockets

The number of these is unlimited but the number of modern telephones or ringing devices (e.g. extension bells) connected at any one time is limited to four. More than this and telephones may not ring or even work.

Cable

The cable used should comply with BT specification CW 1308, which is $1/0.5\,mm^2$ and ranges from four-core (two pairs) to 40 core (20 pairs). It is not usual for secondary sockets to require any more than four cores.

Wiring

Wiring may be flush or surface but kept clear of the low-voltage electrical system by at least 50 mm. No more that 100 m of cable should be used overall and the length between the master socket and the first extension socket should not be more than 50 m.

Connection to the modern insulation displacement connectors (IDCs) terminals is made using a special tool provided with each socket. The connection requirements are as shown in Figure 8.4.

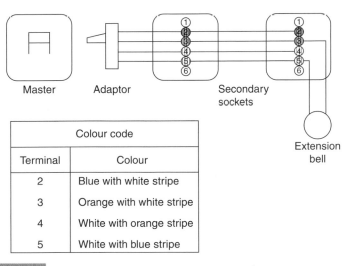

Colour code	
Terminal	Colour
2	Blue with white stripe
3	Orange with white stripe
4	White with orange stripe
5	White with blue stripe

FIGURE 8.4 Telephone socket wiring.

Appendix 1
Basic Electrical Theory Revision

This appendix has been added in order to jog the memory of those who have some electrical background and to offer a basic explanation of theory topics within this book for those relatively new to the subject.

ELECTRICAL QUANTITIES AND UNITS

Quantity	Symbol	Units
Current	I	Ampere (A)
Voltage	V	Volt (V)
Resistance	R	Ohm (Ω)
Power	P	Watt (W)

Current

This is the flow of electrons in a conductor.

Voltage

This is the electrical pressure causing the current to flow.

Resistance

This is the opposition to the flow of current in a conductor determined by its length, cross-sectional area, and temperature.

137

Power

This is the product of current and voltage, hence $P = I \times V$.

Relationship between voltage, current and resistance

Voltage = Current × Resistance $V = I \times R$,
Current = Voltage/Resistance $I = V/R$ or
Resistance = Voltage/Current $R = V/I$

Common multiples of units

Current I amperes	kA Kilo-amperes 1000 amperes	mA Milli-amperes 1/1000 of an ampere
Voltage V volts	kV Kilovolts 1000 volts	mV Millivolts 1/1000 of a volt
Resistance R ohms	MΩ Megohms 1 000 000 ohms	mΩ Milli-ohms 1/1000 of an ohm
Power P watts	MW Megawatt 1 000 000 watts	kW Kilowatt 1000 watts

Resistance in series

These are resistances joined end-to-end in the form of a chain. The total resistance increases as more resistances are added (Figure A1.1).

Hence, if a cable length is increased, its resistance will increase in proportion. For example, a 100 m length of conductor has twice the resistance of a 50 m length of the same diameter.

Resistance in parallel

These are resistances joined like the rungs of a ladder. Here the total resistance decreases the more there are (Figure A1.2).

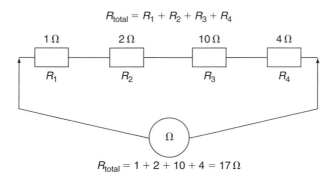

$$R_{total} = R_1 + R_2 + R_3 + R_4$$

$$R_{total} = 1 + 2 + 10 + 4 = 17\,\Omega$$

FIGURE A1.1 Resistances in series.

$$1/R_{total} = 1/R_1 + 1/R_2 + 1/R_3 + 1/R_4$$

$3\,\Omega$

$6\,\Omega$

$8\,\Omega$

$2\,\Omega$

Ω

$$1/R_{total} = 1/R_1 + 1/R_2 + 1/R_3 + 1/R_4$$
$$= 1/3 + 1/6 + 1/8 + 1/2$$
$$= 0.333 + 0.167 + 0.125 + 0.5$$
$$= 1.125$$
$$\therefore R_{total} = 1/1.125$$
$$= 0.89\,\Omega$$

FIGURE A1.2 Resistances in parallel.

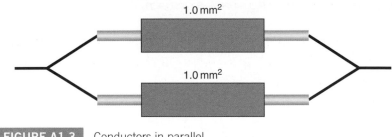

FIGURE A1.3 Conductors in parallel.

The insulation between conductors is in fact countless millions of very high value resistances in parallel. Hence an increase in cable length results in a decrease in insulation resistance. This value is measured in millions of ohms (i.e. megohms, $M\Omega$).

The overall resistance of two or more conductors will also decrease if they are connected in parallel (Figure A1.3).

The total resistance will be half of either one and would be the same as the resistance of a $2.0\,mm^2$ conductor. Hence resistance decreases if conductor cross-sectional area increases.

Example A1.1

If the resistance of a $1.0\,mm^2$ conductor is $19.5\,m\Omega/m$, what would be the resistance of

1. $85\,m$ of $1.0\,mm^2$ conductor
2. $1\,m$ of $6.0\,mm^2$ conductor
3. $25\,m$ of $4.0\,mm^2$ conductor
4. $12\,m$ of $0.75\,mm^2$ conductor?

Answers
1. $1.0\,mm^2$ is $19.5\,m\Omega/m$, so, $85\,m$ would be $19.5 \times 85/1000 = 1.65\,\Omega$
2. A $6.0\,mm^2$ conductor would have a resistance six times less than a $1.0\,mm^2$ conductor, i.e. $19.5/6 = 3.25\,m\Omega$
3. $25\,m$ of $4.0\,mm^2$ would be $\dfrac{(19.5/4) \times 25}{1000} = 0.12\,\Omega$
4. $12\,m$ of $0.75\,mm^2$ would be $19.5\,m\Omega/m \times 1.5 \times 12\,m = 0.351\,\Omega$.

POWER, CURRENT AND VOLTAGE

As we have already seen, at a basic level, power = current × voltage, or $P = I \times V$. However, two other formulae can be produced: $P = I^2 \times R$ and $P = V^2/R$. Here are some examples of how these may be used:

1. A 3 kW 230 V immersion heater has ceased to work although fuses, etc., are all intact. A test using a low-resistance ohmmeter should reveal the heater's resistance, which can be determined from:

   $P = V^2/R$

   So, $R = V^2/P$

   $$= \frac{230 \times 230}{3000} = \frac{52\,900}{3000} = 17.6\,\Omega$$

   This can be compared with the manufacturer's intended resistance. This would show that the element is not broken and further investigation should take place (probably a faulty thermostat).

2. Two lighting points have been wired, incorrectly, in series. The effect on the light output from two 100 W/230 V lamps connected to these points can be shown as follows:

   Each lamp will have a resistance of $R = V^2/P$ (when hot)

   $$= \frac{230 \times 230}{100} = \frac{52\,900}{100} = 529\,\Omega$$

   It will be seen that each lamp will have only 115 V as a supply (Figure A1.4). Hence each will deliver a power of $P = V^2/R$, giving

   $$\frac{115 \times 115}{529} = 25\,\text{W}$$

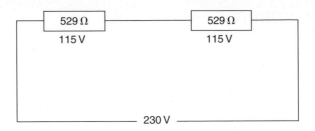

Lamps in series.

which is a quarter of its rated value, and so both lamps will be only a quarter of their intended brightness.

3. The current flowing in a 10 m length of 2.5 mm² twin cable is 12 A. The resistance of such cable is approximately 0.015 Ω/m, so the power consumed by the cable would be:

$$P = I^2 \times R$$
$$= 12 \times 12 \times 0.015 \times 10 = 21.6\,\text{W}$$

Appendix 2
Conductor Identification, Notices and Warning Labels

CONDUCTOR IDENTIFICATION

The following are the colours/alphanumeric references required by the IEE Wiring Regulations:

	Conductor	Letter/Number	Colour
Single-phase AC			
	Line	L	Brown
	Neutral	N	Blue
Three-phase AC			
	Line 1	L1	Brown
	Line 2	L2	Black
	Line 3	L3	Grey
	Neutral	N	Blue
Control wiring or ELV			
	Line	L	Brown, Black, Red, Orange, Yellow, Violet, Grey, White, Pink or Turquoise
	Neutral		Blue
For all systems			
	Protective		Green-yellow

NOTICES AND WARNING LABELS

WARNING
ISOLATE SUPPLY
BEFORE OPENING
COVER

DANGER
400 VOLTS
◀——— BETWEEN ———▶

 CAUTION

THIS INSTALLATION HAS WIRING
COLOURS TO TWO VERSIONS OF BS 7671.
GREAT CARE SHOULD BE TAKEN
BEFORE UNDERTAKING EXTENSION,
ALTERATION OR REPAIR THAT ALL
CONDUCTORS ARE CORRECTLY
IDENTIFIED

I.E.E Approved   to BS 5499-1 2002

WARNING
ELECTRIC
SHOCK RISK

THIS INSTALLATION OR PART OF IT, IS
PROTECTED BY A DEVICE WHICH
AUTOMATICALLY SWITCHES OFF THE SUPPLY IF
AN EARTH FAULT DEVELOPS.

TEST QUARTERLY BY PRESSING THE BUTTON
MARKED "T" OR "TEST".

THE DEVICE SHOULD SWITCH OFF THE SUPPLY,
AND SHOULD THEN BE SWITCHED ON TO
RESTORE THE SUPPLY.

IF THE DEVICE DOES NOT SWITCH OFF THE
SUPPLY WHEN THE BUTTON IS PRESSED, SEEK
EXPERT ADVICE.

Do not switch on

Electrician

Working

**SAFETY ELECTRICAL
CONNECTION
DO NOT REMOVE**

 DANGER
400 VOLTS

Index